Thoughts on the Interpretation of Nature
and Other Philosophical Works

Thoughts on the Interpretation of Nature
and Other Philosophical Works

DENIS DIDEROT

Introduced and annotated
by David Adams
Reader in French
University of Manchester

CLINAMEN PRESS

Clinamen Press Limited
Enterprise House
Whitworth Street West
Manchester M1 5WG

ISBN 1 903083 05 2 (paperback)
ISBN 1 903083 06 0 (clothbound)

1 3 5 7 9 8 6 4 2

Typeset in Adobe Caslon by
Koinonia, Manchester
Printed and bound in Great Britain by
Redwood, Trowbridge, Wiltshire

Contents

Foreword

This volume contains the first complete translation into English of Denis Diderot's *Pensées sur l'interprétation de la Nature*, based on the augmented 1754 edition of the work. It also includes translations of two other works by him: the *Lettre sur les Aveugles* and *Le Rêve de D'Alembert*. These were originally written in 1749 and 1769 respectively, though the latter was not seen in print until 1831.

This edition is intended primarily for English-speaking readers; for convenience, references to works written in other languages are accompanied wherever possible by the titles of the English translations, whether or not these are now in print.

In annotating the *Interpretation*, I have, of course, benefited greatly from the work of other Diderot scholars. I am particularly indebted to two previous editors of the text: Paul Vernière, whose edition of the *Œuvres philosophiques de Diderot* (Paris, Garnier, 1964) provides a great deal of abstruse technical information not available elsewhere; and Jean Varloot, whose notes in volume IX of the *Œuvres complètes de Diderot* (Paris, Hermann, 1981) are a model of aptness and lucidity.

Manchester, August 1999   David Adams

Introduction

Denis Diderot was born in Langres, in Champagne, on 5 October 1713. His origins were humble: of his mother Angélique (1677–1748) we know little, except that he seems to have been fond of her, and that she sent him money in Paris during his student days there. His father, Didier Diderot (1685–1759) was, like his forebears, a cutler, solidly-established and well-respected in the town, though not a man of learning. Didier seems to have been inordinately proud of the young Denis, who was the eldest of his five surviving children, and who showed an intellectual precocity unmatched by any of his siblings. His father and uncle hoped that he would make a career in the church. Indeed, to a greater extent than is often realised, Diderot did display in his early years an interest in religion which, in one way or another, was to remain with him all his life. He was, first, a prize-winning (if somewhat fractious) pupil at the Jesuit college in Langres, and in 1726 he received the tonsure which indicated an intention to take holy orders. He then enrolled, at the age of fifteen or so, at the Collège d'Harcourt in Paris, an avowedly Jansenist establishment which allowed him to take his studies further still. Whatever the vagaries which led him to study in two such mutually-antagonistic institutions,[1] Diderot seems to have retained his religious vocation for some considerable time. Indeed, the master's degree which he obtained from the Sorbonne in 1732 led him, as it led so many other young men of similar background, to specialise in philosophy and theology, with every prospect of enjoying a comfortable living in the church. Only in 1735, after he had completed his studies, did he finally give up the idea of an ecclesiastical career.

His decision is not really surprising. His friend Naigeon tells us that, along with philosophy, Diderot studied Greek and Latin, modern languages, the physical sciences, and mathematics as well.[2] This broad curriculum (which has obvious echoes in the *Thoughts on the Interpretation of Nature*) evidently held more appeal for him than the narrow orthodoxies of theology. Having abandoned the Church, he resolved to follow his own path. Free from the constraints of formal study, he developed a particular interest in mathematics, a skill which enabled him to find casual employment during the 1730s as a tutor to the children of wealthy families. For some years, he led a bohemian existence, all the while deepening his interest in philosophical and mathematical questions, and slowly maturing the independence of mind which was to be the hallmark of all his great works.

In the late 1730s, with no secure source of income, he was obliged to take on whatever work came his way. He later claimed, no doubt truthfully, that he reviewed books for the *Observations sur les écrits modernes* edited by the abbé Desfontaines, and for other journals as well. But such hackwork can scarcely have been well paid, and he augmented such sums as he could earn by undertaking the translation of the *Grecian History* (1707) of the English antiquarian Temple Stanyan. While the *Histoire de Grèce* (1743) is perfectly competent, it excited little interest, and was not reprinted for another thirty years. It shows few signs of being the work of an original mind, and certainly has little bearing on the development of Diderot's own ideas. Even so, one ought not to dismiss it without a second thought, for it did enable him to display his skill as a translator. That skill was manifested more strikingly two years later, in the *Essai sur le mérite et la vertu*, his own, and very personal, version of Shaftesbury's *Inquiry Concerning Virtue and Merit* of 1699. This is his first work with any claim to bear the distinctive mark of his personality, and in a number of important respects it points the way to his future achievements.

Before we look at the *Essai* in some detail, one fundamental point needs to be made. Anyone who comes to the study of Diderot expecting to find in his writings clear evidence of a consistent progression from one set of ideas to another will be disappointed. Critics who have approached his *œuvre* from a purely thematic point of view, giving separate consideration to, for instance, his views on æsthetics, or politics, or religion, have been able to do so only by ignoring the chronology of his writings; they have, of necessity, imposed on them an overall, systematic coherence which they emphatically did not have at the time they were written. Of course, one can trace certain filiations of thought from one work to another; even so, Diderot's writings are best considered as attempts to examine a number of philosophical problems from a variety of points of view which sometimes overlap, modify or contradict one another. It is for this reason that we need to look in some detail at the living genesis of his ideas, and see them in the context of his work, rather than as lifeless extracts preserved in the card-indexes of systematising commentators.

We know little of the background to his involvement in the project to translate Shaftesbury's *Inquiry*; it clearly appealed to him, however, as an opportunity to assert his own views, not merely as a translator, but as an interpreter of Shaftesbury as well. The *Essai* appears to start with the best of intentions: Diderot makes the point in the preface that without a belief in God, there is no virtue, and without virtue there is no happiness. But despite this pious formulation, he allows himself an escape route by stating on the same page that 'virtue is *almost* indivisibly attached to the knowledge of God' (my italics), thus allowing the possibility that the two are not inseparable.

The *Essai* itself confirms that God is not really at the centre of Diderot's moral philosophy. Despite the occasional allusion to Christianity, the references to God are at best of marginal importance. It is clear from the prefatory letter to his Jansenist brother, and from the footnotes which he supplied to his translation, that Diderot had come to accept readily the notion of a secular, rational morality in which justification by faith had no place. Despite his claim that without religion there is no morality, the focus of the *Essai* is that ideas such as virtue, goodness and beauty are explicable in purely human terms. The physical beauty of a creature, for example, is defined according to how well its body is equipped to perform the functions for which nature has intended it. Since he also states that nature is one single entity, this definition of beauty must apply universally, and be as valid for the lowest creatures as it is for man. By implication, Diderot has therefore removed from mankind its dominion over nature, and has silently joined the ranks of the materialists for whom man is merely a part of the natural world, not the special creation of God.

Such consequences are still largely implicit in the *Essai*, but they were to be developed more dramatically in other works which were soon to come from his pen, and which bore more strongly still the mark of his originality. Here again, it would be wrong to think in terms of the simple linear development of his philosophical outlook, for his early career as a writer is a strange blend of the public and the clandestine. During the 1740s, he was involved in the translation of texts which were granted the *privilège* required for works to be published with official approval. At the same time, he was developing his own ideas in books of a quite different kind, which the authorities could not openly tolerate, or which they actively suppressed.

Thus, no sooner had he published (with tacit official recognition) his version of Shaftesbury's *Inquiry* than he was recruited to assist with the translation of the *Medicinal Dictionary* of Robert James. This massive and learned work, which had been recently published in London, appeared in its French guise as the *Dictionnaire de Médecine* (1746-48); it was garlanded with official seals of approval, and bore on its title-page the names of its three translators. Since there is no clue to the contribution which each of them made to the undertaking, it is impossible to say for which parts Diderot was responsible, and therefore to what extent he was involved in the enterprise. Even so, the work of translating the *Dictionary* helped him to become familiar with the technicalities of medicine, and thus broadened even further the wide range of specialist knowledge which he had already acquired by self-study and academic training. To this extent, his involvement in the project was another step towards the intellectual mastery of specialised subjects which he was to display from the late 1740s onwards.

Yet at the same time as he was labouring on what must be regarded as little more than journeyman hackwork (no doubt undertaken mainly in order to feed the wife and two children whom he had acquired by 1746), Diderot was secretly writing his *Pensées philosophiques [Philosophical Thoughts]*. This was a work which brought him not so much fame as notoriety. The very title indicates how readily he had embraced the unorthodox thinking which was increasingly making an impact on the more reflective minds in France at the time, and which we think of as being at the very heart of the Enlightenment. Yet, while Diderot is usually regarded as one of the greatest of the *Philosophes,* it has to be remembered that he and the other members of this coterie did not at first see themselves as a brotherhood: that sense of identification began to emerge only during the 1750s. Nor did they refer to themselves by this name, at least initially. In fact, when the word *Philosophe* first became current, it was as much a term of abuse used by the defenders of religious and political orthodoxy as a boast made by those they criticised.

It would be difficult to argue that all the *Philosophes* believed the same things, or that they evolved consistent, commonly-accepted doctrines on all the questions which preoccupied them. Some were atheists, others were deists of one kind or another; some were republicans, others royalists; some believed in democracy, others distrusted the people; some believed in material progress, others decried it as the worst of human afflictions. But there was at the core of their beliefs a conviction that the Catholic Church was excessively powerful in the State, and that it was a reactionary and repressive influence on every area of intellectual life. Sooner or later, they argued, its power must be broken if France was to slough off the limitations under which her people lived, and from which the citizens of England were so manifestly free. They believed also that knowledge comes via the senses, and that our individual perceptions are what determine our view of the world. Since every man may see the world differently, it was essential to reason about our beliefs, in order to arrive at some agreed position in philosophy, politics, and so forth. They therefore rebelled against the fact that they were forbidden to question religious doctrines which their reason could not accept, and which had nothing to justify them except their allegedly divine origins. These common positions emerged only gradually, and after much hesitation and uncertainty; the Enlightenment, understood in these terms, was a process which occurred over decades, rather than years. Yet such ideas underlay Diderot's thinking from an early stage in his career as a writer, and they surface unmistakably in the *Philosophical Thoughts*.

Though they begin innocuously enough, with some remarks in defence of the passions, the *Thoughts* soon move into the more controversial, and even dangerous, territory of deism and atheism. Despite the apparent assurance of the beliefs he had expressed in the *Essai*, Diderot seems now much less sure of his philosophical and theological views. He seems in some places to incline towards deism (sections XIII, XV, XIX, XX), arguing at one point that the discoveries of physicists and botanists have given proof of the wonders of God's creation (XVIII). Elsewhere, he seems to prefer scepticism in religious matters (XXIV, XXX, XXXIII). Yet there is also a materialist streak in the work, which leads him to argue, for example, that an infinite number of possible combinations of atoms could explain any configuration of matter (XXI); hence, there is no need to suppose the existence of a Creator. The first two-thirds of the book (up to section XL) consist, then, of an internal debate between the conflicting impulses which drove him now in one direction, now in another, as his previous confidence deserted him. But the last third is unmistakably devoted to reducing Christian apologetics to a series of mocking nonsequiturs, in which the Scriptures, miracles and the divinity of the Church are called squarely into question with relentless scorn. If the hand of God is visible in the Scriptures, Diderot asks, why are they so badly written (XLV)? He provides an answer by saying that the truth of the Scriptures matters more than their style, but insinuates that they are clearly the work of human, rather than divine, hands. Nor will he have any truck with miracles, however many witnesses might attest to them (XLVI). Since the truth of a religion should be capable of rational demonstration, its proponents should have no need of miracles (L). Miracles are witnessed, and believed in, by those who are determined to see them (LIII). While Diderot proclaims his submission to the decisions of the Catholic Church (LVIII), his true sympathies clearly lie elsewhere, since belief in Christianity requires faith, which is incompatible with reason (LIX). The authority of the Church derives from its assertion of the divinity of the Scriptures, the truth of which it alone claims the right to interpret. Since that truth cannot be demonstrated rationally, the Church's claim to the obedience of the faithful has no independent foundation (LX). Although the last two sections are an indication of Diderot's perplexity in deciding which kind of religious belief to adopt, he finally gives preference to deism, but without great enthusiasm (LXII).

The *Philosophical Thoughts* offended official sensibilities for several reasons, and caused uproar in orthodox religious circles. Many of the comments which Diderot makes about the Christian tradition, about miracles and about human intervention in the allegedly divine had been made before, notably by writers of the previous century such as Pierre Bayle (1647–1706).[3] But they had chiefly been made in large folio volumes which were expensive, and which were read by few people outside the ranks of professional theologians and philosophers. The *Philosophical Thoughts*, on the other hand, was a small, compact volume of no more than 150 pages, including the index. Its division into short, incisive segments meant that it could be read as and when the reader chose, without the need for prolonged concentration. The danger to the faith (and to the faithful) was obvious. The book was published clandestinely in the spring of 1746, and was condemned by the Paris Parlement, or legislature, on 7 July. The sentence, stipulating that it was to be lacerated and burned by the public hangman, was intended both as a sign of official condemnation, and as a warning to other authors who might be tempted to follow a similar path. As might have been expected, official disapproval did nothing to dampen the book's commercial success. The *Thoughts* were to be Diderot's most successful work to date. It was, indeed, to prove one of the most enduring of all his writings: some thirty reprints, translations and refutations were published during his lifetime, and others appeared even after his death nearly forty years later.

From the standpoint of Diderot's intellectual development, the *Philosophical Thoughts* is a work of major importance. It laid bare his deep dissatisfaction with orthodox religion of any kind, and his refusal to embrace any system of belief which was repugnant to his reason. While deism exerts some appeal (as it had in the *Essai sur le mérite et la vertu*), his tendency now is to suspend belief in any religious system, however abstract, and to prefer scepticism as the only way to preserve his intellectual independence. He had not yet embraced unalloyed atheism, but it is clear that, by 1746, his adherence to systematic religion of any kind was waning fast.

Even so, the condemnation of the *Thoughts* did serve as a warning to him to be careful about what he published. The following year, a manuscript of his next work, *La Promenade du Sceptique*, was seized by the police in his own home, and his repeated requests for its return went unheeded. As its title indicates, the *Promenade* continues the exploration of some of the ideas adumbrated in the *Philosophical Thoughts*, though its ponderous allegorical style would no doubt have proved less attractive to the general reader had it been published in the 1740s.[4] But it does not entirely repeat Diderot's previous views. Most importantly, its affirms a belief in the oneness of the universe, and in the universality of natural laws. Such views had been expressed in the *Essai* too, but less openly, and with a less philosophical emphasis. The atomistic atheist of section XXI of the *Thoughts* is heard no more; in his place is the advocate of the view that the universe obeys some (unspecified) guiding principle, and that creation is identical with its creator. This view Diderot had inherited from Spinoza's *Tractatus* (1677), but he was now to use it in ways which owed little to his predecessor, and much to his own originality of mind.

The *Promenade* is in fact an important stepping-stone between the sceptical phase of Diderot's intellectual development and his wish to find some principle which can explain the unity of creation. As he indicates in section XXI of the *Thoughts*, the achievements of science in discovering both the laws of the macrocosm and the infinite complexity of the microcosm might serve to reinforce belief in the power of God; yet such a conclusion in itself explains nothing unless one can discover what is meant by God, and Diderot readily acknowledges in section XXV of the *Thoughts* that he cannot.

At the same time, since the universe is undeniably governed by the laws of nature, those laws, so Diderot believed, must be explicable rationally. The works written after the *Promenade* constitute a series of attempts to formulate those laws. He does not always attempt to explain the unifying principle of nature in the same way, but his central concern is none the less to discover, if he can, what this principle is. At the same time, he never loses sight of the human dimension in philosophy and science. Of this concern he gives ample proof in a group of works published between 1748 and 1754; these include *Les Bijoux indiscrets*[5] (1748), the *Lettre sur les Aveugles [Letter on the Blind]* (1749) and the *Pensées sur l'Interprétation de la Nature [Thoughts on the Interpretation of Nature]* (1753/4).

Les Bijoux indiscrets, might not seem to qualify as a philosophical text, and indeed there are still critics who refuse to take it seriously. It is a frankly pornographic novel which, using an Oriental setting of the kind popular in fiction at the time, takes up a story found in Classical literature and in medieval *fabliaux*. It concerns the attempts by the sultan Mangogul to find a virtuous woman. Such a woman is defined as one whose sexual organs will not have any scandalous secrets to reveal when the sultan points his magic ring in their direction and causes them to speak the truth. Although Diderot exploits to the full the scabrous opportunities offered by such a scenario, he devotes chapters to the consideration of serious issues as well. In particular, he is at pains to emphasise his dislike of purely systematic philosophy founded on nothing more than theory. His strong preference is for experience and experiment, both of which, if used appropriately, can lead us to the truths of nature in complementary ways. This point is fundamental for an understanding of his later writings, such as the *Thoughts on the Interpretation of Nature,* since it stresses the epistemological importance of the individual observer as well as of the disinterested scientific experimenter. One may doubt, however, that this was the primary reason for the enormous success which the *Bijoux* enjoyed both in France and abroad, with over twenty editions being published within a few years.

Nothing better illustrates the breadth of Diderot's intellectual activities than the other work which he published in 1748, the *Mémoires sur différens [sic] sujets de mathématiques*, the first of his own works to bear his name on the title-page. Diderot was evidently establishing a reliable reputation in Parisian publishing circles: the *Mémoires*, replete with an official *privilège*, were published by none other than Laurent Durand, who had brought out the *Philosophical*

Thoughts only two years earlier, and who had also published *Les Bijoux indiscrets*. None the less, Durand could scarcely have supposed that so abstruse a work as the *Mémoires* would do as well in commercial terms as Diderot's previous (and notorious) efforts, and indeed they did not: the book sold slowly, and was never reprinted separately.

The *Mémoires* mark a change not so much in Diderot's philosophical concerns as in his approach to the problems which had preoccupied him on and off for several years. In the dedication to his mistress Mme de Puisieux, he proclaims that his aim is no longer to mock contemporary foibles, but to devote himself to serious works, and put an end to the scandal surrounding his name. The seriousness of the *Mémoires* is beyond doubt. Yet even in this technical treatise, much of which does not concern us, Diderot remains aware of the human factor. In the first *Mémoire*, which deals with the principles of acoustics, he returns to the idea, expressed in the *Essai*, that the quality of an organ such as the ear is to be judged by its fitness to perform its natural functions. But he goes further than this, arguing that our appreciation of musical sounds is governed by the conjunction of the acoustic laws of harmony, the state of our individual organs, and our own experience of similar sounds in the past. That is to say, scientific understanding requires us to take into consideration not only of the fixed laws of nature, which account for the proportions of musical sounds relative to one another, but also of the experience accumulated by the individual percipient mind.

This latter consideration was Diderot's starting-point in his next work, the *Letter on the Blind*, published in 1749. The *Letter* is divided essentially into two parts. The first consists of a report on a visit made by the anonymous writer and others to a blind man living at Puiseaux (near Orléans). He is said to be capable of such feats as threading a needle unaided, and to have a philosophical turn of mind which enables him to understand much about the physical world despite his handicap.

The main part of this section is devoted to Diderot's own reflections on the differences between the perceptions of the blind and the sighted. Basing himself on the belief that our morality and our view of the world depend to a large extent on the state of our senses and organs, he is certain that the blind have a sense of morality and of religion different from that of the sighted. Stealing is especially abhorrent to them, in that it can be effected secretly. In contrast, nakedness, with its attendant notions of modesty and morality, is of no consequence to those who cannot see it. The wonders of nature, on which so many apologists for religion base their claims, has little appeal for the blind. The situation of a blind man whose sight is restored, and who tries vainly to make other blind people understand what he has seen, is comparable, Diderot says, to that of those rare individuals who have seen the truth but who are persecuted by the defenders of religious orthodoxy. The applicability of this assertion to his own case needs no emphasis. Not unnaturally, he stresses the importance of touch in giving the blind an idea of shape, and it is, he claims, through the reiterated experience of sensations of touch that the blind learn to distinguish one shape from another. The 'interior sense' of the blind man combines these sensations to produce the idea of different surfaces and solids, and imagination in the blind consists of recombining the memory of tactile sensations in different ways.

Much of what we find in this part of the *Letter* is symptomatic of wider changes of outlook during the previous fifty years or so, which are attested in the growth of experimental science, and in the sensory empirical epistemology set out in John Locke's *Essay Concerning Human Understanding* of 1691. Some traces of Lockian epistemology can be found in Diderot's earlier work, but the *Letter* bears stronger traces of this outlook than anything he had previously written. Locke (1632-1704) had revived and developed the old Aristotelian thesis that there is nothing in the mind except what is derived from the senses. It is the mind's ability to compare repeated sensations which gives us true knowledge of external causes, says Locke, who thereby places sensory experience at the very heart of his philosophy. In thus emphasising the importance of sensory perceptions, he provided an alternative to the deductive methods of Cartesian philosophy. René Descartes (1596–1650) had based his conviction that we could know the truth on his belief in innate ideas (such as extension and solidity); he also had massive faith in the deductive power of reason, on which he placed greater reliance than on experiment or experience. Innate ideas and reason he combined with an assertion of the goodness of God who, he argued, would not deceive us as to the truth of our perceptions. Despite its initial appeal, Cartesian physics and metaphysics had proved less and less satisfying to Continental minds in the years since his death in 1650. This was mainly because the experimental results obtained by scientists were increasingly at variance with Descartes' mathematical predictions, and increasingly in line with the results predicted by the work of Isaac Newton. This practical verification of experimentation (which was underlined by the growing international reputation of the Royal Society), was contemporary with Locke's insistence on the centrality of individual experience. These consonant developments, without which the Enlightenment would have been inconceivable, explain to a great extent why Locke's *Essay* (which was translated into French as early as 1700) had an even greater impact on the Continent than in his native country.[6]

In the *Letter on the Blind*, Diderot gives a particular twist to Locke's sensualism by asking us to imagine how a man born without sight would come to understand the world, and what form that understanding would take. Having dwelt on the differences between the perceptions of the blind and the sighted, he turns to the question of how best to achieve meaningful communication between the two. His discussion of this point takes up much of the second part of the *Letter*, which is centred on the alleged experiences of the blind Cambridge mathematician Nicholas Saunderson (1682–1739).

What lies behind much of this section of the work is Diderot's desire to combat the idealism of bishop Berkeley, to which he had fleetingly referred in *La Promenade du Sceptique*. George Berkeley (1685–1753) had argued, in his *Dialogues of Hylas and Philonous* (1713) that we can know nothing beyond our own sensations, and indeed beyond our own existence. There is no guarantee that objects continue to exist when we do not apprehend them by our senses, unless we suppose that God is always there to witness what we cannot experience. In the *Letter*, Diderot says that this solipsistic doctrine (which mixes elements of Lockian empiricism with a Cartesian belief that God guarantees the truth of our perceptions) is 'the most difficult of all systems to combat, even though it is the most absurd'.

By bringing Saunderson into the *Letter*, therefore, he is attempting first and foremost to construct a means whereby the truth can be guaranteed without undue reliance on subjective perception. The first question to which Diderot turns in discussing Saunderson is that of allowing the blind to acquire the notion of number. Saunderson gets over this difficulty by the use of large and small pins arranged in different ways in holes made in a board. Each digit from one to zero is represented by a specific arrangement of pins. By this means, having gradually become familiar with his system, he is able to perform complex calculations at a speed which astonishes the sighted. This mathematical empiricism attempts to circumvent the scepticism of Berkeley's dialogues by moving the focus of human perception from sight to touch (Saunderson is said to 'see' through his skin). Hence,

Diderot is attempting to guarantee the truth of sensory perceptions in a way not specifically addressed by the idealists.

In what was to prove a highly controversial part of the work, Diderot invents a dialogue between the atheistic Saunderson and a minister of the Anglican church called Gervaise Holmes, who attempts to convert him to Christianity on his death-bed. Like the blind man of Puiseaux, Saunderson remains unconvinced by the argument that the wonders of nature proclaim the existence of God. In perhaps the most famous lines in the *Letter*, he tells Holmes: 'If you want me to believe in God, you must make me touch Him.' Nor is he willing to believe in God simply because great men such as Newton have accepted His existence. He goes on to argue (recalling the atheist of section XXI of the *Philosophical Thoughts*) that any combination of molecules is theoretically possible, that nature has always created monsters which could not survive, and that even the survival of the human race may have been no more than a matter of chance. In these circumstances, it is idle to speak of God's plan for mankind. With his last breath, Saunderson asks the God of Newton and of the theologian Samuel Clark to pity him, and dies.

The final section of the *Letter* is a somewhat long-winded examination of what has become known as Molyneux's problem; that is, would a man born blind who recovers his sight be able at once to distinguish a cube from a sphere by simply looking at them? Diderot reviews the answers to this problem given by Locke in his *Essay* and by his friend Condillac in his *Essai sur l'origine des connaissances humaines [Essay on the Origins of Human Knowledge]* (1746). He concludes by asserting that such a man would need to look repeatedly at the differences between the two shapes before being able to distinguish them visually. It is experience, rather than the power of the eyes alone, which will assist him in this task.

The real significance of the *Letter* is that it represents a change of philosophical orientation for Diderot. In place of the uncertainties of the *Philosophical Thoughts*, he now attempts to find some principle which will allow the blind (and the sighted too) to make positive sense of the physical world. This principle is number: Saunderson's calculating pins are a means whereby he can reliably convert the world of unseen phenomena into mathematical terms. Number can be made to operate consistently, and can be perceived accurately, through constant repetition, by our sense of touch.

In much the same spirit as in the *Mémoires sur différens sujets de mathématiques*, Diderot is proclaiming his faith in the power of mathematics to solve philosophical problems. The very fact that the structure of natural objects is ephemeral and unpredictable (as Saunderson's comments on monsters indicate) shows the difficulty of attempting to find any single principle which will explain the workings of external nature. Instead, Diderot turns his attention to the workings of the individual percipient mind, and finds in mathematics the means whereby it can reduce disparate phenomena to the unity of numbers. It needs to be emphasised, too, that the idea of numbers makes sense only if the mind can, through repetition, perceive their consistency. This is why, as in *Les Bijoux indiscrets*, Diderot is at pains to stress the importance of repeated experience, both in his presentation of Saunderson's system and in his discussion of Molyneux's problem.

The *Letter on the Blind* was published (again by Durand) on 9 June 1749. On 24 July, Diderot was arrested at his home and taken to the prison at Vincennes, south-east of Paris. He remained there for over three months. Whether he was imprisoned because of the *Letter* alone, or because he had offended the authorities through the accumulation of impious works written over several years, it is impossible to say. He had, fortuitously, sent a copy of the *Letter* to Voltaire on its publication. While the great man had not been overly impressed by the arguments of the work, he recognised the brilliance of its author, and was sympathetic to his plight in Vincennes. Diderot might well have languished there in dire circumstances had it not been for the intervention of Voltaire's mistress, Mme Du Châtelet, who was a relative of the governor's. While she could not obtain Diderot's release, she was at least able to ensure that he was well treated, and able to take exercise in the grounds of the prison.

Despite his having friends in high philosophical circles, Diderot's stay in prison might well have been longer had he not been associated with the consortium of publishers created to bring out the *Encyclopédie*. He had first become involved with the project in 1746, and from at least 1748 had been its co-editor, with the mathematician d'Alembert. The quantities of time and money already invested in the enterprise were considerable and, without Diderot, might well be wasted. The publishers (who included Durand and the king's printer Le Breton) therefore interceded repeatedly with the authorities for his freedom. They were eventually successful, and he was released on 3 November 1749.

The reasons for his imprisonment and release matter less, however, than the effects of the episode on Diderot himself. In Vincennes, he wrote abjectly to the police, apologising for having given offence in his writings; these letters demonstrate that the experience of imprisonment frightened him, and left a lasting mark on his life and career. From this time onwards, except when he could be sure of powerful protectors, he published little which could offend the spiritual and temporal authorities, and he kept virtually all his most original works under lock and key until his death. Some, indeed, were not published until well into the nineteenth century.

A large part of his time in the next twenty years was taken up with editing the *Encyclopédie*. This was a gigantic task, and his responsibilities as editor make it all the more astonishing that he found any time to publish works of his own. His *Lettre sur les Sourds et Muets [Letter on the Deaf and Dumb]* (1751) exists in no fewer than five separate states, all published within a few months of one another, and all containing revisions and additions which testify to the pressure under which he was undoubtedly working. This second *Letter* develops the investigation of the sensorially-challenged which had begun with the *Letter on the Blind*, but it is a far more complex, convoluted, and even obscure, piece than the first. Essentially, Diderot argues that all languages are built on a common foundation, and that what is nowadays called their 'deep structure' always follows the model of subject-verb-object. Even inflected languages such as Latin can be seen to behave in this way, if we analyse them closely enough. It follows that the structure of the mind is universally the same, and that the perceptions of all human beings are converted into speech in essentially the same way.

This belief in the universal structure of the mind is apparent also in the *Suite de l'Apologie de l'abbé de Prades* of 1752. The young abbé de Prades had contributed theological articles to the *Encyclopédie* which aroused little comment, but his doctoral thesis was condemned by the Sorbonne for its defence of Lockian empiricism. Seizing their chance to mount an assault on the whole enterprise, the Jesuits secured the suspension of the *Encyclopédie* in January 1752. Publication was resumed in October of that year, thanks largely to the support of the more enlightened members of the government (and of the king's mistress Madame de Pompadour). Buoyed up by this expression of confidence, Diderot took the opportunity set forth his ideas in the *Suite* which is printed after De Prades' own defence of his views.

In defending the abbé (and hence the theological opinions expressed in the *Encyclopédie* too), Diderot insists that we have an instinct to believe in the existence of the natural world; we trust our senses to give us true information about it, and to provide evidence of an organising intelligence, whom we call God. The mathematical system of Saunderson is nowhere in evidence, and the wonders of the universe are now invoked for theological purposes which had been explicitly rejected in the *Letter on the Blind*. Following on from the *Letter on the Deaf and Dumb*, Diderot states unequivocally his belief that all human beings (at least all those with the usual five senses) perceive the world in essentially the same way. Hence, the special concerns of the disabled are now laid aside in favour of a 'common sense' Lockian epistemology. The *Suite* looks back to the some of the concerns of the *Philosophical Thoughts* and the two *Letters*. It is the last of Diderot's writings in which belief in the truth of our sense perceptions is given a central and fundamental emphasis. In his subsequent works, this principle is taken for granted, and the uses of empiricism are examined in a more pragmatic light. This is the case with the *Pensées sur l'Interprétation de la Nature* [*Thoughts on the Interpretation of Nature*].

The first edition of this work, entitled simply *De l'Interprétation de la Nature*, was printed towards the end of 1753. Its rarity inclines one to believe that this was a trial issue only, and that it was not intended for public circulation in this form.[7] The book as we know it today only appeared, with tacit permission, early in 1754.[8] As with the *Letter on the Deaf and Dumb*, Diderot revised and added considerably to the initial version of the work (giving a text of 241 pages, as against 204 in the original).[9] To judge from the eccentric pagination, the sometimes careless syntax, and other signs of haste,[10] the revisions must have been undertaken at great speed, and printed as they stood.

In changing the title of the revised version to *Pensées sur l'Interprétation de la Nature*, Diderot was drawing attention to two major features of the work. In the first place, he lays emphasis on the fragmentary character of the book, and no doubt alludes to his earlier *Pensées* of 1746, which were still proscribed. This is understandable, since it represents what might be called 'work in progress' rather than a finished treatise. It is a product of his reflections at that time, informed by the considerable study which he had undertaken in recent years, particularly in connection with the *Encyclopédie*. In the second place, in stating that nature can be interpreted, Diderot is voicing his optimistic conviction that it speaks a language, that it is therefore essentially coherent, and that it has an underlying structure which we can understand. He makes clear his belief that, if the human mind is capable of intelligent investigation, and if individuals are capable of collaborative effort, then the laws of nature should to some degree be intelligible to percipient minds working together for the same ends.

The *Interpretation* is addressed to a young audience, as the prefatory note makes clear. From the outset, Diderot explicitly rejects mathematics as a source of useful information about nature (sections II, IV). He proclaims instead his faith in 'l'expérience' (a word which, as in *Les Bijoux indiscrets*, can mean 'experience', though it usually needs to be understood here as 'experiment'). Using the experimental approach to test his hypotheses (what would now be called 'the scientific method'), the young scientist will explore the marvels of nature, with due humility, and with acknowledgement of his own limitations (VI–X, XXII). Much will depend on his recognising that nature is a unity, that it is the product of what Diderot calls 'a single act', and that even one natural phenomenon incompatible with the laws of nature would undermine the whole of philosophy (XI). This belief is the basis of the hypothesis that nature originally created only one living creature, from which all others are descended, through innumerable variants (XII). It therefore follows that seminal fluid is produced by females as well as by males (XIII).

Having developed the idea of the unity of nature, Diderot then goes on to assert that it can best be understood through an (admittedly rare) combination of observation, reflection, imagination and experimentation (XV, XXV–XXXI), which eschews the blinkered approach of the purely rational philosopher (XXIII–XXXI). The examples of the conjectures which such a mind may engender (XXXII–XXXVIII) range widely, and include the nature and causes of uterine molæ; the differences between the processes of art and those of nature, and ways to improve the quality of the steel made in France.

It is the duty of the philosopher, Diderot claims, to ensure that his discoveries are expressed in ways intelligible to those who are not experts (XXXIX, XLI). He also has a duty to test his hypotheses in every conceivable way (XLII–XLIV): it is his task to perform as many experiments as he can in order to identify the true nature of the phenomenon under investigation; scientists need to fill in the gaps separating these apparently disparate phenomena and demonstrate, if possible, their unity (XLV–XLVI).

The next three sections (XLVII-XLIX) which concern the dangers of an over-systematic approach to investigation, are really a prelude to section L, which is in effect a long review of Maupertuis' *Dissertatio [...] de universali systemata naturæ* (1751). Maupertuis (who published the book under the pseudonym of 'Baumann', which Diderot uses here) is reproached with not having thought through the consequences of his hypothesis that the formation of animals can be explained by supposing that molecules experience desire, aversion, memory and intelligence. If all molecules do not behave in this way, says Diderot, then there is no unity in the universe, and the laws of physics, and indeed the laws of God, break down into incoherence. If all molecules do behave like this, then 'Baumann' is in effect suggesting that the universe is one great animal, and creation is identical with its creator, God. Diderot is implying that 'Baumann' is in fact advocating the same doctrine as Spinoza, and is thus a heretic (a conclusion which did not appeal to the pious, though frequently inebriated, Maupertuis). Instead of supposing the molecules have so many active properties, Diderot suggests, 'Baumann' should have endowed them merely with the capacity to arrange themselves automatically in the most convenient formation, rather as animals unconsciously change their position while they are asleep (LI).

The following section of the work (LII–LVII) consists of a series of exhortations to investigators to ensure that they are properly prepared for their task; they should have adequate instruments; they should be conscious of the importance of their task, and should be able to move by solid hypothesis from what is known, measurable and certain to what still remains to be discovered. But let them remember, too, that they are interpreters of nature; their task is to explain the nature of phenomena; they should not attempt to determine the final cause of things, for that is unknowable. Lastly, the investigator should not fall victim to the common error that there is nothing new under the sun, for such folk-wisdom is the enemy of impartial investigation.

In section LVIII, entitled 'Questions', Diderot sets out what is in some ways the most significant part of the work: he takes up the conjectures for which he had pretended to mock Maupertuis, and states his view that matter is indeed a unity, and operates according to universal laws. At the same time, species may develop slowly, over millions of years, perhaps, passing through different stages of viability and completeness, so that species which (as faithful Christians) we take to be fixed and eternal may only appear to be so. Similarly, what we take to be distinct states of matter, some of which seem to be living and others dead, may in fact be merely different stages of its development. Hence, he suggests, any species, whether animal, vegetable or mineral, may alter its composition radically over time, as its current molecular content changes with accretion or loss, and it becomes something else. It may become sensitive at a particular stage of its development, just as it may subsequently lose this property. So, he asks, is living matter always living? or dead matter always dead? How is the passage from one state to another achieved? To these questions, which Diderot himself calls 'futiles' in relation to the practical needs of human beings, he has no answer; it is enough for him to have suggested the problems to which the next generation of investigators can turn their attention.

Even for a modern-day reader, the *Interpretation* is a somewhat puzzling work. Despite being addressed ostensibly to the young researcher, it requires an expertise in a number of different scientific fields which few young readers were likely to possess. Nor is it really a primer of scientific method, since it suggests the kind of questions which the experimenter should investigate, but in many cases gives few indications of how the experiments should be conducted. It contains a number of conjectures, but little indication of how they might be verified. At a practical level, therefore, the *Interpretation* seems to offer less than might have been hoped.

Yet the real importance of the work lies elsewhere. It is, historically, nothing less than a thoroughgoing, vigorous defence of the scientific method, and a rejection of the rationalist, predictive approach to science inspired by Descartes and practised by his followers. In particular, Diderot has little time for the systematic, mathematical philosophy dear to the rationalist school, which he accuses of being preoccupied with theorising and unconcerned with experimental facts. Though he does not name d'Alembert, his co-editor of the *Encyclopédie* (who was by no means a pure Cartesian), his strong criticisms of mathematics as ultimately futile and divorced from reality can be read as evidence of their growing estrangement at a personal as well as professional level.[11]

If the *Interpretation* represents a rejection of mathematics (and thus of the central doctrine of the *Letter on the Blind*), it constitutes a major step towards fostering the scientific mentality which we recognise as fundamental to the shaping of the modern world. In addition, one ought not to mistake the importance of its emphasis on conjecture, a term which invites readers to allow their thoughts to roam where they will, in the hope of discovering new connections in nature, or new avenues of investigation. Nothing could be further from the rigidities of formal rational thought.[12] It is essential to understand also that the work was abreast of the latest scientific thinking. As he acknowledges in his own notes to the text, Diderot was stimulated in his ideas on the nature of matter by the recently-published work of scientists such as Buffon and Maupertuis. A number of other contemporary sources of information are not specified, but are readily traceable. In particular, many of the technical details provided on physics, chemistry and geology can be found in articles composed by a variety of contributors for the first four volumes of the *Encyclopédie*. There is thus no doubt that Diderot made good use of his position as editor to keep himself informed of the latest developments in those fields. While the *Interpretation* obviously could not rival the depth and scope of the *Encyclopédie*, he was clearly setting out to distil into it much that was most novel and significant in the scientific articles of the larger work. It could thus serve both as a digest for those who had no access to the *Encylopédie* itself, and as a summary of his ambitions for that work which would stand even if its enemies should one day prevail, and secure its abandonment before it could be published in full.

While clear affiliations between the *Interpretation* and articles in the *Encyclopédie* can certainly be traced, they also share, at a more fundamental and diffuse level, the same source of philosophical inspiration. Far from representing the latest thinking, it harks back a century and more to the work of the English philosopher Francis Bacon (1561–1626). Both the *Prospectus* for the *Encyclopédie* (1750), and the 'Système figuré des connaissances humaines' which forms part of the preliminaries of volume I (1751), are explicitly based on Bacon's work. The 'Observations' accompanying the 'Système figuré' reveal likewise a detailed knowledge of his scientific writings. The *Preliminary Discourse* in the same volume (which was written mainly by d'Alembert, with Diderot's participation) describes Bacon as 'immortal', and as perhaps the greatest of all philosophers.

The editors were particularly indebted to *The Advancement and Proficiencie of Learning* (1603) and its more elaborate second version, the *Novum Organum* of 1620. Much of the intellectual driving-force behind the *Encyclopédie* comes essentially from these two seventeenth-century English works.[13] The editors shared Bacon's vaunting ambition to bring within their compass all branches of human activity, all aspects of nature, and ultimately therefore the whole of creation. In attempting to accomplish their purposes, they adopted the Baconian perspective, in which all branches of knowledge were ultimately related, like those of a tree, and the central, unifying trunk of that knowledge was the human mind. To this end, like Bacon himself, they relegated 'Memory' and 'Imagination' to the margins of their system, and made 'Philosophy' the major category of human intellectual endeavour, embracing all the reasoning faculties of the mind, and all the sciences. At the same time, they did not copy Bacon slavishly: they refused to see ecclesiastical history as central to history itself, and claimed the right to judge all religions on rational criteria, a view which Bacon had regarded as impious; and they gave natural history an importance which Bacon, no doubt echoing the attitudes of his time, did not.

The article 'Baconisme, ou philosophie de Bacon' (1752) gives further clues to the way in which the English philosopher's ideas were perceived in the *Encyclopédie*. Although it bears the signature 'G' used by the abbé Mallet, Diderot, as editor, was aware of its contents, and clearly took the same view of Bacon's importance as Mallet himself. 'Baconisme' might serve, in fact, as a summary of the main theses put forward both in the *Encyclopédie* and in the *Interpretation*. Mallet stresses that Bacon had liberated philosophy from the shackles of scholasticism, which had placed vacuous verbal theorising above the direct investigation of nature; if science was to progress, it needed a new kind of approach, a new understanding of the world of nature, and this Bacon had provided. True scientific method, Mallet reports him as saying, moves forward from experiment to provisional conclusions or axioms, and thence by slow degrees to more general statements about nature. The true philosopher will use the widest range of experiments, as well as his reflection and judgement, and be as free as possible from the prejudices of his times.

The *Interpretation* starts from these basic Baconian principles, though it does not follow them unswervingly. What it offers, essentially, is a translation of the major aspects of Bacon's thought, rewritten and repackaged into Enlightenment terms. Like his predecessor, Diderot concerns himself greatly with the general laws of nature as they can be learned by induction. Yet he is more interested than Bacon in the pragmatic analysis and understanding of specific phenomena, such as the effect of electricity on crystals, or the improvement of French steel. There is no discussion in the *Interpretation* of such non-scientific matters as popularity and loquacity which are analysed in the *Advancement*. Nor is Diderot greatly concerned with the moral philosophy of that work, where the idea of goodness is analysed at length. Finally, as might be expected of the editor of the *Encyclopédie*, there is, if anything, more emphasis in Diderot than in Bacon on science as a collaborative enterprise involving men of all persuasions and of widely differing abilities.

Hence, there are undeniable changes of emphasis between the *Advancement* and the *Interpretation*. None the less, if one looks at the background to the *Interpretation*, it is clear that Diderot intended it to be nothing less than an eighteenth-century equivalent of the *Novum Organum*; it was thus invested, for the initiated at least, with all the scientific and historical importance which the *Encyclopédie* had already given to the earlier work. Indeed, as some readers were well aware, the very title of the *Thoughts on the Interpretation of Nature* was a direct reminder of the *Interpretatio naturæ*, a short treatise which Bacon had incorporated in to Book II of the *Novum Organum*.

But well-informed, literate readers did not make their voices heard very loudly when the *Interpretation* was published early in 1754. Although its qualities were praised here and there in articles written for the *Encyclopédie*, the general reaction was one of bafflement, and even derision. The enemies of the *Philosophes* seized on it avidly as proof that here was another dangerous independent thinker, whose incomprehensible jargon masked an ambition to overthrow all established notions of rationalism, and to replace orthodox belief with deism, or even atheism. This hostility and mockery may well explain why the *Interpretation* was the last philosophical book which Diderot published for many years. Indeed, its reception marked the beginning of a period of disappointment for him as an author. He published two plays later in the decade: *Le Fils naturel, ou l'Epreuve de la Vertu [Dorval, or the Test of Virtue]* in 1757, and *Le Père de Famille [The Father of a Family]* in 1758. Despite their originality, and his attempts to move French theatre in the direction of greater realism with the *drame bourgeois*, they met with little immediate success, and he published nothing further for the theatre during the remainder of his life.[14]

Nor could he hope for much consolation in editing the *Encyclopédie* which, in the late 1750s, ran into even greater difficulties with the authorities than earlier in the decade.[15] As a result, d'Alembert's participation in the work was much reduced after 1758. With little effective support, Diderot had then to soldier on with the massive task of editing the work, to which also he continued to contribute a large number of articles. When it was condemned by Church and State in 1759, and its *privilège* withdrawn, it seemed as though years of the most intensive effort had come to nothing. It took some time for matters to become easier. The publishers were eventually permitted to bring out the volumes of plates (the first of which appeared in 1762), but it was not until 1772, after yet further obstacles had been overcome, that the whole enterprise was brought to a successful, if exhausting, conclusion under Diderot's direction.

These various setbacks had shown unmistakably that, faced with combined and concerted forces of Church and State, he could count on few supporters even among his own colleagues. And he had not forgotten his time in Vincennes a decade earlier. For understandable reasons, therefore, he was unwilling to risk his own exposed position as editor of the *Encyclopédie* by hinting to anyone at the nature of the thoughts which he committed to paper over the next decade or so. To a degree, he limited his public appearances, so to speak, by addressing himself to the very select readership of the *Correspondance littéraire*. This was a manuscript newsletter, edited by his friend Grimm, which was sent at fortnightly intervals to a dozen or so royal or aristocratic subscribers in foreign countries who wished to keep up with the news from Paris. Diderot contributed to it articles and reviews, and later substantial works as well; a large part of his most original output from the mid-1750s onwards (particularly the art-criticism of the *Salons*) was therefore seen first by this very small and restricted foreign audience.

Yet there were other, no less significant, works whose existence was understandably not divulged to the readers of the *Correspondance*, or even to Diderot's intimates. Two in particular stand out. The first was his self-questioning satirical dialogue *Le Neveu de Rameau [Rameau's Nephew]*. This was written mainly between 1762 and 1772, but did not appear in an accurate French version until 1823. The other, *Le Rêve de D'Alembert [D'Alembert's Dream]*, dates from 1769, but did not see the light of day until 1831.

Both works are complex (the *Nephew* particularly so), and range far beyond purely philosophical topics. Even so, they can properly be regarded, in part at least, as developments of the ideas expressed in the *Interpretation* and earlier works. In the *Interpretation*, Diderot had supposed that matter could at some stage become endowed with sensitivity. He is now concerned only with matter which already possesses sensitivity, and in the *Nephew* he proceeds to explore this hylozoist hypothesis from the point of view of its effects on the individual. The *Nephew* (designated 'He') argues that, if we are simply animated matter, an essentially random collection of molecules, then we have no control over the constitution of our bodies, and hence no responsibility for our actions. If that is so, then punishment is an irrelevance, and justice has no meaning. This view is countered by the narrator ('Myself') who believes in virtue and self-control, and in the deterrent effect of punishment. Diderot had been preoccupied since the mid-1750s with the question of whether and how to punish wrongdoers, and no firm answer to this dilemma is forthcoming in the *Nephew*. The work is far more important for its explorations of Diderot's doubts and uncertainties than for his ability to resolve them; yet this would hardly have served as a defence for a work which, in debating such issues, also attacked the morals and motives of many of the leading figures in contemporary Parisian society.

Firmer philosophical answers are, however, forthcoming in *D'Alembert's Dream*. The work is divided into three sections: the conversation between Diderot and d'Alembert alone, the *Dream* itself (which involves d'Alembert, his mistress Mlle de L'Espinasse, and Doctor Bordeu), and the subsequent discussion between the last two of these three. The initial conversation is based on the belief that all matter possesses sensitivity. But this view, which is advanced only tentatively in the *Interpretation*, is now developed as a positive, and indeed verifiable, basis for the explanation of the unity of nature.

All matter is endowed with the capacity for movement; the motion of bodies is proof of its presence, but even the transition of molecules from the inert to the active state is due to the inherent presence of movement in matter. The distinction between animate and inanimate nature is merely the distinction between two accretions of molecules at different stages of their existence (the 'futile' questions relating to 'dead' and 'living' matter raised in the last section of the *Interpretation* have now been answered). The inanimate can become animate as, for example, when we change the nature of food by digesting it. Similarly, the act of conception (such as that of d'Alembert himself, who was an illegitimate foundling) is nothing more than the conflation of two sets of molecules provided by the parents (who have existed for all eternity in other kinds of matter); only the form, not the nature, of the molecules changes in the process. Matter has the capacity not only to feel, but also to remember, and even to form ideas. Memory is nothing more than the fact of one fibre of the brain striking another, rather like the keyboard of a harpsichord; this causes a particular vibration in the brain, and thus produces a specific association of ideas. Human beings are, in Diderot's words, 'instruments endowed with sensitivity and memory', and sensitivity is 'a general property of matter'. Since all matter is ultimately one, there is no need to suppose the existence of a soul to animate the body, or to believe that any form of matter can never be alive.

Despite the sometimes playful tone of their exchanges (understandable in a work written for private recreation on Diderot's part), the issues involved are serious. At the heart of this section is an attack on the 'mind-body' dualism of Descartes, as important in its way as the attack on Cartesian rationalist science in the *Interpretation*. Cartesianism was founded on the thesis that the mind and body are essentially separate in nature and essence, and that contact between the two (via the pineal gland, so Descartes alleged) was essentially a process of communication between utterly distinct entities. This theory, which had gained official acceptance in French educational circles by the middle of the eighteenth century, was Diderot's target in the *Dream*. In discussing the development of a chicken from the embryo in the egg to a living creature, he concludes that if, as Descartes would have it, the chicken is a machine, then man too is a machine, since both are made of the same matter. This is to remove the special status of human beings, not only in Cartesian terms, but in religious terms as well, and to reject the claims of theology to pronounce on the nature of man. The purpose of this first section, then, is to lay down the (very perilous) materialist foundations on which the rest of the work will be constructed.

The second section, d'Alembert's dream itself, teases out the consequences of these ideas. By using the literary convention of the dream (a device which goes back to Classical times), Diderot is free to make his speaker pursue the most outlandish and extravagant train of thought without being considered mad, except by Mlle de L'Espinasse. But with Dr Bordeu on hand to explain the meaning of the dream, she is eventually reassured.

The central concern of this section is the question of individual identity. If we are all parts of the same matter, then our individuality is at best a fleeting one, and we may become incorporated into a quite different species as our molecules become once more part of the matter from which they originated. Sensitivity is inherent in living molecules, and becomes common to both when two molecules join together, rather like bees in a swarm. Indeed, sensitivity may continue to exist even when the swarm is divided, so that the limits of our individuality are blurred and uncertain.

We do of course have a continuing sense of our own identity, through our memories, and we regard our brains as the centre of our being, like spiders at the centre of their webs. The brain is the meeting-point of all the filaments which allow it to communicate with every part of the body, and thus the brain is the seat of consciousness. Since each part of the body harbours memories of its own particular type of experience, all of which are recorded in the brain, identity is a question of retaining some aggregate of these individual memories. Such a notion of identity may, even so, be wholly illusory *sub specie æternitatis*. At the level of mankind in general, on what we would now call the diachronic plane, it would be wrong to suppose that human beings have always been, or will always remain, as they are now. In a changed environment, their bodily needs would adapt to changed circumstances. Even on the synchronic plane, there is no certainty that any human being will be born with all the 'normal' features we expect. Molecules accrete to produce bundles of filaments, each of which usually develops in order to perform a particular task, such as giving us our five senses. Yet (as Diderot had argued in *Rameau's Nephew*) any unpredictable alteration to the conformation of a molecule could give rise to a filament which produces a hitherto-unknown organ or sense, or could lead to any number of 'monstrous' distortions of the body. Indeed, it may only be our social conventions which prevent us from realising that the male is a 'monstrous' version of the female, and vice versa.

Our sense of our own being is likewise a fragile one: accidents can rob us of our memories, and hence of our sense of our own identity, or bring about changes to our behaviour over which we have little control. Our constitution and character, too, may be dictated by our physical conformation. If the 'branches' or bundles of filaments which connect with our centre of consciousness are less vigorous than the 'trunk' from which they all radiate, the individual in question will be characterised by strong imagination and emotions. A properly balanced system, in contrast, will produce thinkers and philosophers. There may also be men who struggle to overcome the effects of excessive sensitivity, and who are constantly torn between their physiological nature and their desire to master it. Such men will remain calm amidst great dangers, and it is from their ranks that great figures of every kind emerge.[16] In sleep, the filaments themselves may become agitated and send a message to the centre, or the reverse may occur. The relationship between them when consciousness is in abeyance is therefore unpredictable, and this is why dreams can be so distinct from our waking experiences, even though they may be as real to us. The intellectual audacity and sweep of this section are probably without parallel in the literature of the Enlightenment. Certainly, there is nothing as strongly argued or as wide-ranging in Diderot's other writings on the same theme. There are points which can be traced back incidentally to ideas put forward in Buffon's *Natural History* (which had begun publication in 1749, and was directly relevant to the *Interpretation*). None the less, in its daring and originality, the *Dream* represents the furthest development of Diderot's own materialist ideas, rather than those of his contemporaries. Indeed, there are times when he seems almost to anticipate such modern discoveries as the role of DNA in genetics, and the significance of physiology in determining behaviour.

Even without crediting Diderot with the gift of prophecy, we can see that he is impatient with the intellectual constructs of his own time. He does away with Cartesian dualism, and makes the soul a purely material construct, since it is nothing more, he argues, than the centre of our consciousness. We have no eternal existence, except as molecules of matter. We are the products of our physiological conformation, and our nature is the product not of original sin, but of our molecules, about which we can do very little. Monsters are the sign not of depravity or our sinful nature, but of a molecular malformation. All our actions, whether awake or asleep, have mechanical causes, and are explicable in physiological terms alone. Though the question is not raised in these terms in the *Dream* , we can hear in the background the voice of Rameau's Nephew asserting that to punish wrongdoers for their physiological conformation is both pointless and ineffectual.

These speculations are rounded off by a very short coda, the *Suite de l'Entretien [The Outcome of the Conversation]*. Since we are impelled to propagate our species, Bordeu argues, religious celibacy should be condemned as harmful to the individual and to society; pursing this train of thought, he defends masturbation and even homosexuality as providing harmless relief, though he would not say so publicly. Nor does he shy away from the possible advantages of mixing the species, arguing that bestiality could produce a race of goat-men whose could serve as footmen without the loss of dignity suffered by human beings who take on such menial tasks.

In addition to the hostile circumstances besetting the *Encyclopédie*, there were sound personal reasons why Diderot kept these writings to himself, and in manuscript form only. In the *Nephew* and in the *Dream*, he takes the liberty of depicting friends and acquaintances in ways which they would certainly not have welcomed. Jean-François Rameau, the insignificant nephew of the composer, was only a marginal member of Diderot's circle; more substantial figures from Parisian society who are mocked in the *Nephew* would not have taken lightly to having their intimate lives set so accurately and fully before the public by a man they detested. D'Alembert, on the other hand, had been a colleague and collaborator on the *Encyclopédie* for over twenty years. His mistress Mlle de L'Espinasse and his friend Bordeu were long-standing acquaintances of Diderot's.

When she somehow discovered his secret, Mlle de l'Espinasse made known her intense distaste at being portrayed as a woman deeply interested in unusual sexual behaviour. D'Alembert, who was outraged, demanded that work be destroyed forthwith. Furious as their reactions were, Diderot was fortunate to have provoked the anger only of his friends and acquaintances. Had the existence of the *Dream* been suspected by the authorities, official disapproval would certainly have made itself felt with considerably greater force. The ideas expressed in the final section would have been enough on their own to ensure the rapid and unequivocal condemnation of the work. To criticise religious celibacy was to fly in the face of the laws and traditional teachings of the Church. To defend masturbation was to oppose not only the Church, but the virtually unanimous wisdom of the medical profession as well. To defend homosexuality was to set oneself against laws both human and divine, since in eighteenth-century France men who practised it were committing a serious, and even capital, crime.[17] To defend bestiality was to put oneself beyond the pale, and offend absolutely every responsible agency in society.

But there is a double veil here: not only does Diderot not publish the *Dream*; he also allows Bordeu to utter views which the doctor himself says that he would not express in public. That is to say, there are some thoughts which may be too dangerous to put into every mind; some things are better left unsaid, or at least said only in private, for fear of the harm that may be caused.

It would not perhaps be going too far to see in this self-imposed silence of Bordeu's the key to Diderot's own public reticence in later years. From the 1770s onwards, he wrote as assiduously as ever, but few of his writings were offered to the public during his lifetime. To the last dozen or so years of his life belong (to mention only the most important works) his major political writings, such as his forceful (but anonymous) contributions to the abbé Raynal's *Histoire des deux Indes* *[History of the East and West Indies]*; his *Entretiens avec Catherine II* *[Conversations with Catherine II]* and other radical texts written as a result of his visit to Russia in 1773–74; his *Essai sur Sénèque,* and fictional texts such as the short stories and the novel *Jacques le fataliste* *[Jacques the fatalist]*. Apart from his contributions to Raynal's *History* (after 1770) and the life of Seneca (1779), these texts remained unpublished until long after his death in 1784 (though a few did appear in the *Correspondance littéraire*). When they finally became available in print, they met with an indifference and an incomprehension which was to last, almost without a break, until the twentieth century. Only in our own time have Diderot's genius and originality come to be appreciated as they deserve, and the process of re-assessment is still far from complete. The works printed in this volume bear witness to the breadth and depth of his interests, and to the vigour and originality of his thought. The reader who approaches them with an open mind, free of the prejudices against which Diderot rails in so many of his writings, will be amply rewarded.

Notes

1 For much of the eighteenth century, the Jesuits and the Jansenists were at loggerheads over the central doctrinal questions of predestination and sacramental grace. The Jesuits were accused of being too ready to grant absolution for sins, without ensuring that the recipient was truly repentant; in turn, the Jesuits acused the Jansenists of relying on the doctrine that God's grace was bestowed for reasons unknowable to man, and of having excessve faith in predestination.

2 *Mémoires historiques et philosophiques sur la vie et les ouvrages de D. Diderot* (Paris, 1821), p.5.

3 Bayle's *Dictionnaire historique et critique [Historical and Critical Dictionary]* was first published in 1697. It was frequently revised and augmented, and went through a number of editions until the early nineteenth century.

4 It did not in fact become available until 1831, in a volume of hitherto-unpublished works by Diderot which also contained the *Paradoxe sur le comédien* and *Le Rêve de D'Alembert.*

5 It was translated into English in 1749 as *The Indiscreet Toys.*

6 See Ross Hutchinson, *Locke in France 1688–1734, Studies on Voltaire and the Eighteenth Century,* 290 (Oxford, 1991).

7 Only two copies are known. One is in a private collection; the other (which surfaced at auction in Paris as recently as 1995) is in the Bibliothèque nationale de France, with the accession no. Rés. p. R. 1062. A similar 'trial issue' of the *Pensées philosophiques* also exists.

8 Two other editions were published in French in 1754, but they derive from the revised version used here, and have no particular textual significance.

9 There are in addition eleven unnumbered pages of index.

10 In particular, Diderot refers in section XLIII to statements which he had made in the initial version of section XXXIII, but had subsequently deleted. As a result, the reader can make no sense of the reference.

11 Their relationship was under strain for several reasons. Following the suspension of the *Encyclopédie*, d'Alembert had written to Voltaire on 24 May 1752, threatening to concern himself with the mathematical part of the work only, thus depriving Diderot of his support, Diderot may also have been jealous of d'Alembert: his *Mémoires sur différens sujets de mathématiques* clearly did not have as much originality as the mathematical works of his co-editor, who had also shown his versatility by publishing a translation of extracts from Tacitus in 1753. It is also true that the two men had widely different notions of the epistemology of knowledge: d'Alembert wished to fit every branch of enquiry into an over-arching metaphysical structure, an approach which was at odds with Diderot's more flexible and empirical view of research. To judge from d'Alembert's praise of the *Pensées sur l'Interprétation de la Nature* in the *Encyclopédie*, he did not take personal offence at Diderot's attack on mathematics. On their editorial relationship, see Véronique Le Ru, 'L'aigle à deux têtes de l'*Encyclopédie*: accords et divergences de Diderot et de d'Alembert de 1751 à 1759', *Recherches sur Diderot et sur l'Encyclopédie*, 26 (April 1999), 17–26.

12 Diderot's epigraph to *Rameau's Nephew* (see below) is 'My thoughts are my whores'; one might say as much for the role of conjecture in the *Interpretation*.

13 It should not be forgotten that the *Encyclopédie* had its origins in another English text: the project initially devised by the publishers envisaged the translation into French the *Cyclopedia* of Ephraim Chambers.

14 His ideas on the drama were, however, to be taken up later on in France and, especially, in Germany, where the plays and the theoretical writings which accompanied them were translated by Lessing. See Roland Mortier, *Diderot en Allemagne* (Paris, 1954). Diderot wrote other plays, as well as the *Paradoxe sur le Comédien* (1770–73), but they remained unpublished until the nineteenth century.

15 For the details, see Arthur M. Wilson, *Diderot* (New York, 1972), chapter 21.

16 In the *Paradoxe sur le Comédien [The Paradox of Acting]*, Diderot applies this theory to the case of the great actor, who must remain inwardly calm even when depicting the most intense emotions.

17 The *Encyclopédie* article 'Sodomie', published in 1765, points out that, in 1750, two men found guilty of having had a homosexual relationship had been burned alive on the Place de Grève in Paris. Diderot's reflections on lesbianism (which was not a crime) are to be found in his novel *La Religieuse [The Nun]*, written c. 1760 and published in 1796.

Suggestions for further reading

The study of Diderot's life and work, and of the Enlightenment itself, has burgeoned internationally in recent years, and it is impossible to list here more than a very few of the many books and articles in English devoted to the subject. Readers in search of critical help with the texts printed in this volume should consult F. W. Spear's *Bibliographie de Diderot* (2 vols, Geneva: Droz, 1980–88) which provides an admirably detailed list of studies in all languages. Those who wish to keep up with new research should consult the *Recherches sur Diderot et sur l'Encyclopédie*. This is published (usually on an annual basis) by the Société Diderot in Langres, and is available in many academic libraries.

A. On the Enlightenment

Cassirer, Ernst: *The Philosophy of the Enlightenment* (Princeton University Press, 1951)

Charlton, D. G: *New Images of the natural in France* (Cambridge: Cambridge University Press, 1984)

Crocker, L. G: *An Age of Crisis : Man and World in Eighteenth-Century French Thought* (Baltimore: the Johns Hopkins Press, 1959)

Gay, Peter: *The Enlightenment, an Interpretation* (2 vols, London: Weidenfeld and Nicholson, 1967)

Hazard, Paul: *European Thought in the Eighteenth Century* (Harmondsworth: Penguin Books, 1965)

Sweetman, John: *The Enlightenment and the Age of Revolution 1700–1850* (London: Longman, 1998)

B. On Diderot

Diderot Studies (Geneva, Droz, 1949–)

Adams, D.J., *Diderot, Dialogue and Debate* (Liverpool: Francis Cairns, 1986)

Bremner, Geoffrey: *Order and Change: the Pattern of Diderot's Thought* (Cambridge: Cambridge University Press, 1983)

France, Peter: *Diderot* (Oxford: Oxford University Press, 1983)

Furbank, P.N: *Diderot, a Critical Biography* (London: Secker and Warburg, 1992)

Lough, John: *The* Encyclopédie *of Diderot and D'Alembert* (Cambridge: Cambridge University Press, 1954)

Lough, John: *The* Encyclopédie *in eighteenth-century England, and other essays* (Newcastle upon Tyne: Oriel Press, 1970)

Mason, John Hope: *The Irresistible Diderot* (London: Quartet Books, 1982)

Morley, John: *Diderot and the Encyclopedists* (2 vols, London: Macmillan, 1878)

Wilson, Arthur M: *Diderot* (New York: Oxford University Press, 1972)

Thoughts on the
Interpretation of Nature

To young persons preparing to study Natural Philosophy[1]

Young man, open this book and read on. If you can manage to reach the end, better books than this will not be beyond you. Since my purpose is not so much to instruct you as to exercise your mind, it matters little to me whether you adopt or reject my ideas, so long as you give them your full attention. Someone more able than myself will teach you how to become acquainted with the power of nature; I shall be content if I have helped you to try out your own powers. And so farewell.

PS. One more word before I take my leave. Always bear in mind that *nature* is not *God*, that a *man* is not a *machine* and that a *hypothesis* is not a *fact*; you may be sure that if you think you have found something here which conflicts with these principles, you will have failed to understand me.[2]

Thoughts on the
Interpretation of Nature

Quæ sunt in luce tuemur
E tenebris.
Lucretius, *Liber VI.*[3]

I. Nature is to be my theme. I shall let my thoughts flow from my pen in the order in which things occur to me, to give a better picture of the workings of my mind. These thoughts will consist either of general views on the experimental method, or of particular views on a phenomenon which appears to worry all philosophers these days, and to divide them into two categories. One category consists, it seems to me, of those who have a great many instruments but few ideas; in the other are those who have a great many ideas but no instruments. The interests of truth would best be served if the intellectually inclined finally deigned to associate with their more active colleagues; speculative spirits would then have no need to actually do anything, and those who toil ceaselessly with their hands would do so to some purpose; all our efforts would be united and directed at the same time against the resistance of nature; and in what might be called this league of philosophers, everyone would play the part which suits him best.[4]

II. One of the truths which have been most boldly and forcefully[*] announced in our own time – one which a good physicist will always bear in mind, and one which will certainly lead to the most beneficial results - is that the domain of mathematicians is a world purely of the intellect, where what are taken for absolute truths cease entirely to be so when applied to the world we live in.[6] This has led to the con-

[*] See the *Histoire naturelle, générale et particulière*, Vol. I, *Discours I* (Note by Diderot)[5]

clusion that it was the task of experimental philosophy to rectify geometrical calculations, and the logic of this view has been acknowledged even by geometricians themselves. But what is the use of correcting geometrical calculations by experiment? Would it not be quicker to take note of the result of an experiment? It is therefore clear that, without experimentation, mathematics, with its essentially transcendental approach, leads to nothing precise; it amounts to a kind of generalised metaphysics,[7] in which bodies are stripped of their individual qualities; and there would still be a need for someone to write a great work with the title *The Application of experimentation to geometry* or *A Treatise on the aberration of measurements.*

III. I do not know whether there is any connection between having an aptitude for gambling and having a mathematical turn of mind; but games of chance and mathematics have a great deal in common. Setting aside the uncertainty of fate on the one hand, or comparing it with the inaccuracy of abstraction on the other, a game of chance can be considered as an indeterminate series of problems to be resolved according to a certain set of conditions. There is no aspect of mathematics to which this definition is inapplicable; and a mathematician's ideas have no greater reality in nature than those of the gambler. Both are a matter of convention. When geometricians decried metaphysicians, they never dreamt that their entire discipline was nothing more than metaphysics. Someone once asked: 'What is a metaphysician?' A geometrician replied: 'A man who doesn't know anything.' It seems to me that chemists, physicists, natural scientists and all those devoted to the experimental method, who are no less adventurous in their judgements, are about to avenge metaphysics and to define geometricians in the same way. 'What, they say, is the good of all these profound theories about the heavenly bodies, all these vast calculations dealing with rational astronomy,[8] if they do not save Bradley or Le Monnier[9] the trouble of observing the heavens?' And I say: 'A geometrician is fortunate if his absorbtion in the study of the abstract sciences has not blunted his appetite for the fine arts; if he is as familiar with Horace and Tacitus as with Newton; if he can discover the properties of a curve and enjoy the beauties of a poet; if his mind and works will never date, and if he is honoured by every academy!' He will never fall into obscurity, and need not fear out-living his fame.[10]

IV. We are at the dawn of a great revolution in science. To judge from the inclination men's minds would appear to have for ethics, literature, natural history and experimental physics, I would almost go so far as to assert that, within the next hundred years, there will hardly be three great geometricians in Europe. This branch of science will just cease at the point where Bernoulli, Euler, Maupertius, Clairaut, Fontaine and d'Alembert have left it.[11] It will stand like the Pillars of Hercules and no-one will pass beyond.[12] Their works will endure in centuries to come, like the Egyptian pyramids, massive and laden with hieroglyphics, an awesome picture of the might and resources of the men who built them.

V. At the dawning of any new science, all men's minds are of course drawn to it as a result of the high esteem in which society holds inventors, the desire to familiarise oneself with something which is much discussed, the hope of distinguishing oneself by some discovery, and the ambition to share the honours with men of worth. Any new science is instantly cultivated by many people of different types. These may be worldly individuals, who find it difficult to live with their own idleness, or fickle minds aspiring to build themselves a reputation in a science which is fashionable, having already failed to do so in other branches of science which they have now abandoned; some make a career of it and others take it up out of inclination. So much concerted effort combines to make this branch of science achieve its potential quite quickly. But as its boundaries are stretched ever further, the esteem in which it is held diminishes. Only the most outstanding exponents are accorded any consideration. And then the crowd thins out; no-one sets off then for a land where fortunes are rarer and harder to make. Science is then left only with mercenaries who earn a crust from it, and a few men of genius to whom it continues to bring honour long after its prestige has faded, and eyes have been opened to the futility of their work. Their accomplishments are still regarded as *tours de force* which do honour to mankind. That, in brief, is the history of geometry, and of every branch of science when it ceases to instruct or to delight. Even natural history is no exception.[13]

VI. When we compare the infinite number of phenomena in nature with the limitations of our own intelligence and the frailty of our organs,[14] how could we ever expect to discover – in view of the slowness of our work, the long and frequent interruptions which it suffers, and the scarcity of creative spirits – anything but a few broken, isolated parts of the great chain which links everything together.[15] Even if experimental science continued to work for century after century, the materials which it accumulated would eventually have become too great to fit into any system, and the inventory of them would still be far from complete. How many volumes would be needed to encompass just the terms intended to designate different sets of phenomena, once these phenomena had been ascertained? How long will it take for the language of philosophy to be complete? And, even if it were complete, what man could possibly master it? And if, as an even clearer sign of his omnipotence than the wonders of nature, the Almighty had deigned to sketch out the mechanisms of the universe on sheets inscribed in his own hand, is this great tome likely to be more intelligible to us than the universe itself? And how many of its pages could have been understood by the sage who, even with all his great intellectual powers, was not sure that he had grasped the chain of reasoning which had led a geometrician in the past to determine the relationship between a sphere and a cylinder? For us, these pages would represent a fairly good yardstick of our breadth of mind, but an even better satire on our vanity. We could say: Fermat reached this page or that,[16] and Archimedes went a few pages further. But what is our purpose? To create a work which can never be accomplished, and which would be far beyond man's intelligence if it were ever completed. Are we not even more insane than the first inhabitants of the plain of Sennar?[17] We know the infinite distance between earth and the heavens, and yet we never tire of trying to raise the tower. But can we assume that the time will never come when we pocket our pride and abandon the project? Does it seem likely that, housed as he is in cramped and uncomfortable conditions here on earth, man would persist in building a palace in which he could live beyond the atmosphere? And even if he did persist, would he not be halted by the babble of languages, which is already all too noticeable and too troublesome in natural history?[18] It is also true that the idea of 'usefulness' sets boundaries on everything. The criterion of usefulness is about to place limits on geometry, and in a few centuries from now, it will do the same for experimental science. I estimate that this field of study will last for some centuries yet, because it has an infinitely broader spectrum of use than any abstract science, and because it is indisputably the basis of everything which we know for certain.

VII. So long as something exists only in the mind, it remains there as an opinion, or a notion which may be either true or false, and which can be accepted or contradicted. It becomes meaningful only when linked to things which are external to it. This linkage is achieved either by an uninterrupted series of experiments, or by an uninterrupted line of reasoning, one end of which is rooted in observation and the other in experimentation;[19] or else by a series of experiments scattered at intervals in a reasoned argument, as weights may be attached along a thread held by its two ends. Without these weights, the thread would be at the mercy of the slightest breath of air.

VIII. Concepts which have no foundation in nature may be compared to those Northern forests where the trees have no roots. It needs nothing more than a gust of wind, or some trivial event, to bring down a whole forest of trees – and of ideas.[20]

IX. Men have scarcely begun to realise how rigorous are the laws governing enquiry into the truth, and how few are the means at our disposal. The whole enterprise comes down to proceeding from the senses to reflection, and from reflection back to the senses: an endless process of withdrawing into oneself, and re-emerging. This is how bees work. We will have foraged in vain if we do not return to the hive loaded with beeswax. All this wax will have been accumulated in vain, unless we know how to make honeycombs.

X. Unfortunately, however, it is quicker and easier to commune with oneself than to consult nature. That is why reason tends to remain cloistered, whereas instinct wants to reach outside itself. Instinct never ceases to watch, to sample, to touch and to listen; there may be more experimental science to be learnt from studying animals than by following the courses given by a professor. There is no artifice in what they do. They set about achieving their purposes, careless of what is around them; if they do take us by surprise, that is not their intention. Astonishment is the first reaction to any great phenomenon; it is the task of philosophy to dispel it. The purpose of a course in experimental philosophy is to send the listener away better informed, not stunned. To pride oneself on natural phenomena, as though one had invented them oneself, would be to imitate the stupidity of an editor of the *Essais*, who could never hear the name of Montaigne without blushing.[21] There is a fundamental lesson which there is often occasion to teach:

the recognition of one's own inadequacy. Would it not be better to win the confidence of others by frankly admitting 'I simply do not know', than to keep babbling on and cover oneself with embarrassment by endeavouring to find explanations for everything? Anyone who openly admits his ignorance of something he knows nothing about makes me more inclined to believe what he does try to explain to me.[22]

XI. Astonishment often arises from imagining several extraordinary events when only one has taken place, and from imagining as many discrete occurrences in nature as there are phenomena, whereas perhaps there has never been more than a single act of nature. It further appears that if nature had been obliged to produce more than one such act, the differing results of these acts would have remained separate; moreover, there would be sets of phenomena unrelated to one another, and the common connecting chain, which philosophy takes to be continuous, would be broken at several points. The total separateness of an individual fact is incompatible with the concept of a whole, and without that philosophy would cease to exist.[23]

XII. It would appear that nature has chosen to use the same mechanism in an infinite number of different ways.* She never abandons one type of creation before replicating that genus in all its possible variations. If we consider the animal kingdom, and observe that, among the quadrupeds, every single one possesses functions and bodily parts - especially internal organs - fully resembling those of any other quadruped, is it not easy to believe that in the beginning there was only a single animal which served as prototype for all the others, and that all nature has done is to lengthen, shorten, alter, multiply or eliminate certain organs?[24] Imagine the fingers of the hand joined together, with the substance of the nails so extended and thickened that it engulfs and covers the whole body; then, instead of a human hand, you would have a horse's hoof.† When we observe the successive outward metamorphoses which take place in this prototype, whatever it may be, pushing one realm of life closer to another by

---

* See *Histoire naturelle*, Tom. IV, *Histoire du Cheval*, and a little work written in Latin with the title *Dissertatio inauguralis metaphysica, de universali Naturæ systemate, pro gradu Doctoris habita*, printed in Erlangen in 1751, and brought to France by Mr de M*** in 1753 (*Note by Diderot*).

† See *Hist. nat. gén et part.* Tom. IV. *Description du Cheval* by Mr Daubenton (*Note by Diderot*).

imperceptible stages, and populating the regions where these two realms border on each other (if they can be referred to as 'borders' in the absence of any true divisions); and, populating, as I said, the border regions of the two realms with vague, unidentifiable beings, largely devoid of the forms, qualities and functions of one region and assuming the forms, qualities and functions of the other; who, then, would not be persuaded that there had never been more than one single prototype for every being? But whether one accepts this philosophical conjecture as true, like Doctor Baumann.[25] or rejects it as false, in common with Monsieur de Buffon, no-one will deny that it should be adopted as an essential hypothesis for the advancement of experimental physics, of rationalist philosophy and for the discovery and explanation of phenomena which depend on being organised. Obviously, nature could never have preserved such a degree of similarity amongst its constituent parts, and introduced such variety in the forms it adopts, without frequently bringing out something in one organism which has been suppressed in another. In this, nature resembles a woman who likes to dress up, and whose different disguises, exposing first one part of herself and then another, give some hope to her ardent admirers that they may one day get to know the whole person.[26]

XIII. It has been found that the same seminal fluid exists in both sexes.[27] The parts of the body containing this fluid are now known. The singular alterations which occur in certain female organs when the woman is hard pressed by nature to seek a male have been noted.* In the coming together of the sexes, when one compares the signs of pleasure on either side and has ascertained, by clear and characteristic spasms, that rapture is achieved on both sides, one could never doubt that comparable emissions of seminal fluid have taken place. But where and how does this emission take place in a woman? What becomes of the fluid? What pathway does it take? This will be known only when nature itself, which is not always so mysterious in all places and at all times, reveals itself in another species. This will probably happen in one of the following two ways: either the shape of these organs will become more pronounced, or else emission of the fluid will manifest itself at its place of origin, and all along its pathway, by its remarkable abundance.[28] Anything clearly seen in one being will

---

* See the *Histoire naturelle,* and especially the *Discours sur la génération* (*Note by Diderot*).

not fail to become apparent in another being of a similar type. The empirical approach to the physical sciences teaches us to identify lesser examples in greater phenomena, just as the rationalist approach teaches us to recognise greater organisms from lesser examples.

XIV. I picture the vast realm of the sciences as an immense landscape scattered with patches of dark and light. The goal towards which we must work is either to extend the boundaries of the patches of light, or to increase their number. One of these tasks falls to the creative genius; the other requires a sort of sagacity combined with perfectionism.

XV. We have three approaches at our disposal: the observation of nature, reflection and experimentation. Observation serves to assemble the data, reflection to synthesise them and experimentation to test the results of this synthesis. The observation of nature must be assiduous, just as reflection must be profound, and experimentation accurate. These three approaches are rarely found together, which explains why creative geniuses are so rare.[29]

XVI. The philosopher often grasps the truth in just the same way as an inept politician, noticing an opportunity, asserts that, like the side of a head which has no hair, it simply cannot be grasped, at the very moment when the experimenter's hand chances upon the side which has some hair. Yet it has to be admitted that, amongst these experimenters, some are very unfortunate: one might devote a lifetime to observing insects without seeing anything new, while another, casting a passing glance at them, might discover the polyp or the hermaphrodite aphid.[30]

XVII. Has the universe lacked men of genius? Not at all. Is the problem then their failure to study and reflect? Even less so. The history of science is studded with famous names; the earth's surface teems with memorials to our work. Why, then, do we have so little certain knowledge? Why has it been the destiny of science to make so little progress? Are we fated always to be nothing but children? I have already given my answer to these questions. The abstract sciences have too long – and too fruitlessly – occupied the best minds; either we have failed to study what needed to be known; or else we have failed to apply selectivity, opinions or method to our studies. Words have proliferated endlessly and the knowledge of things has lagged behind.[31]

XVIII. The right approach to philosophy, both in the past and now, would have been to apply the understanding to what has been understood; to apply understanding and experimentation to the senses; the senses to nature, and nature to the investigation of the instruments to be used; and finally, to employ these instruments for researching into and perfecting the arts, which should be set before the public to teach people to respect philosophy.

XIX. There is only one way of truly recommending philosophy in the eyes of the common people, and that is by associating it with usefulness.[32] The common man always asks the question: 'What's the good of that?' and one should never find oneself in the position of having to answer 'None.' He does not realise that what gives enlightenment to the philosopher and what serves the common man are two very different things, since the philosopher's mind is often illuminated by something harmful and clouded by something useful.[33]

XX. Facts themselves, of whatever type, are the philosopher's true wealth. But one of the abiding convictions of rationalist philosophical thought is that anyone incapable of counting his coins will hardly be any richer than the man who has only a single coin.[34] Unfortunately, the rationalist school of thought is much more concerned with drawing together and connecting the facts it possesses than in accumulating new ones.

XXI. Collecting facts and making associations between them are two very onerous occupations, so philosophers have divided these two tasks up between them. Some spend their lives collecting materials, labouring hard and usefully; others – vainglorious architects as they are – hasten to make use of their work. But virtually all the edifices of rationalist philosophy have been overturned with the passage of time. Sooner or later the dusty labourer will drag up, from the depths where he has been blindly delving, the piece which will fatally undermine the structure erected by force of intellect; it will crumble, leaving only a haphazard mass of rubble, until such time as another bold spirit undertakes some new combination of materials. The systematic philosopher is fortunate if, like Epicurus, Lucretius, Aristotle and Plato in the past, he is endowed by nature with a vivid imagination, great eloquence and the skill to express his ideas with striking and sublime imagery! The structure he has erected may well collapse one day; but his statue will remain standing amongst the ruins and a stone breaking away from the hillside will not shatter it, because it does not have feet of clay.[35]

XXII. The intellect has its preconceptions just as the senses have their own uncertainty, the memory its limits, the imagination its pale flickerings, and instruments their shortcomings.[36] Phenomena are endless; their causes are hidden and their pattern is, perhaps, transitory. All we have for dealing with such obstacles, both those within us, and those imposed by nature from without, is the slow accumulation of experience and our limited powers of reflection. Such are the levers with which philosophy intends to make the earth move.[37]

XXIII. We have identified two types of philosophy – one is empirical and the other rationalist. One of the two goes blindfolded, always groping its way, grasping everything which comes to hand and finally encountering precious things. The other assembles these precious materials and attempts to fashion them into a flaming torch; but this would-be torch has, until now, served less well than the gropings of the rival camp - and this is as it should be. Experimentation moves endlessly, and is forever active; it devotes as much time to seeking out phenomena as reason spends on seeking analogies. Experimental science does not know what its work will produce and what it will not, but it nonetheless labours without respite. Rationalist philosophy, in contrast, weighs up the alternatives, pronounces on them and stops there. It boldly states that *'light cannot be split'*; meanwhile the experimental philosopher merely listens without rejoinder throughout the centuries and then, suddenly, he brings out the prism, with the words *'light can be split'*.[38]

XXIV. Here is an outline of experimental science.[39]

In general, experimental science deals with the existence, the properties and the use of the objects it studies.

EXISTENCE encompasses their *past history, description, generation, preservation* and *destruction*.

Their *past history* concerns their location, the process of importing and exporting them, their cost, the preconceptions surrounding them, etc.

Their *description*, both internal and external, includes all observable properties.

Generation moves from their earliest origins up to the state of perfection.

Preservation uses every means of maintaining them in that state.

Destruction includes everything between the state of perfection and the last known point of *decomposition* or *disintegration, dissolution* or *reconstitution*.

QUALITIES may be general or individual.

Those which I call *general* are common to all things and only vary in quantity.

Those which I call *individual* are qualities which make a thing what it is; they apply either to the whole being, or else to one of its parts or to its decomposed state.

The concept of USE encompasses *comparison, application* and *combination*.

Comparison is based on resemblance or differences.

Application should be as wide and varied as possible.

Combination is either analogous or odd.

XXV. I use the words *analogous* or *odd* because everything produces its effect in nature, from the most extravagant to the best planned of experiments. Experimental science sets out without preconceptions and is always happy with whatever comes along, whereas the rationalist school is always well-prepared with ideas, even when its suppositions are not borne out.

XXVI. Experimental science is an innocent field of study, and scarcely requires the soul to prepare itself at all.[40] The same cannot be said of other schools of philosophy, the majority of which promote an appetite for conjecture which experimental philosophy eventually suppresses. Sooner or later, the appetite for wild conjecture begins to pall.

XXVII. A taste for observation may be instilled into anyone, although a taste for experimentation, it would seem, should be instilled only in those who are well off.

Observation requires nothing more than the normal use of the senses, whereas experimentation demands continual expenditure. It would be desirable for men of rank to add this ruinous pursuit to the many other – less respectable – ones they have devised. All in all, it would be better for them to be impoverished by a chemist than stripped of everything by tradesmen; better to be engrossed in scientific experiment, which would afford occasional amusement, than stirred by the shadow of an elusive pleasure, which they never cease to pursue. I would readily give the same advice to thinkers of limited means with an inclination for scientific experiment as I would to a friend of mine if he were tempted to enjoy a beautiful courtesan: *Laïdem habeto, dummodo te Laïs non habeat.*[41] And I would give the same advice to anyone with sufficient breadth of mind to imagine systems, and sufficiently wealthy to be able to test them by experimentation: Yes, by all means have a system, but do not let yourself be ruled by it: *Laïdem habeto.*

XXVIII. The good effects of scientific experiment may be compared to the advice of a father whose dying words to his children were that a treasure was buried in his field, but he did not know exactly where. His children set about digging the field; they did not find the treasure they sought, but they reaped a rich harvest that season which they had not expected.[42]

XXIX. The following year, one of the men's children said to his brothers: 'I have carefully inspected the land which our father left to us, and I believe I have discovered the site of the treasure. Listen: this is how I have worked it out. If the treasure is buried in the field, there must be certain signs within it which mark the place; well, I have noticed some unusual markings near the corner which faces east: the ground looks as though it has been disturbed. We have already made sure, through the work we did last year, that the treasure is not to be found near the surface; therefore it must be buried deeper. Let us take up our spades here and now and dig until we come to the miser's underground hoard.' The brothers all set to work, carried away less by the force of reason than by a desire for riches. They had already dug deep without finding anything; their hopes were beginning to fade and mutterings were heard. Then, one of them, seeing some shining fragments, formed the idea that he had come across a mine. It was indeed a mine – a lead mine which had been exploited in the past. They worked the mine, which gave them a plentiful yield. The experiments suggested by the observations and systematising notions of the rationalist school of thought sometimes produce a similar outcome. That is how chemists and geometricians, obsessed with solving problems which are perhaps insoluble, came to make discoveries more important than the solution itself.

XXX. The long-established habit of conducting experiments gives even the lowliest practitioners a feeling of knowing what is about to happen which is akin to inspiration.[43] It is entirely up to them whether they make the same mistake as Socrates – of calling it their 'familiar spirit'. Socrates was so unusually experienced at assessing men and weighing up circumstances, that even in the most delicate of situations an accurate and rapid process occurred secretly within him that was followed by a prediction which was never belied by events. His judgement of individuals was based on feeling, in the same way as men of taste judge works of the intellect. The same applies to experimental science and to the instinct of our great practitioners. They have seen nature at work so frequently, and from so close at hand, that they can fairly accurately guess at her likely course, even if they decide to provoke her by embarking on the oddest type of trials. The greatest service they can perform for those they initiate into empiricism is, therefore, not so much to introduce them to procedures and results as to imbue them with a propensity for divination, enabling them to 'scent', so to speak, unknown procedures, fresh experiments and hitherto-undiscovered results.

XXXI. How is this propensity passed on? Anyone endowed with it should look within himself to get a clear picture of it, replace the 'familiar spirit' by clear and intelligible concepts, and then develop them for the benefit of others. Should he should find, for instance, that it is 'a facility for supposing or perceiving contrasts or analogies which is rooted in a practical knowledge of the physical properties of subjects taken in isolation, or of their reciprocal effects when taken in combination'; he would extend this idea, supporting it with innumerable facts occurring to his memory. This would be a faithful account of all the apparently extravagant thoughts which have run through his mind. I use the word 'extravagant' – for how else could one describe such a sequence of conjectures, founded on contrasts and resemblances so remote and so imperceptible that the dreams of a sick man appear neither more strange nor more disjointed in comparison?[44]

Sometimes there is not a single proposition that cannot be disputed, either as it stands or in relation to what precedes or follows it. It forms such a precarious whole, both in its suppositions and in its consequences, that it has often been rejected as a basis for observations or experiments.

XXXII. *Conjectures, first series*

1. There is a certain body known as a mola.[45] Some maintain that this singular body is engendered in the female without the assistance of the male. However the mystery of generation may be accomplished, both sexes are certainly involved. Could the mola not be an assembly, either of every element emanating from the female in the production of a male, or of all the elements emanating from the male in his different approaches to the female? Could these elements which are quiescent in the male, but widespread and sustained in certain females with a hot temperament and a vivid imagination, not be fired and stirred into activity? And could those elements which are quiescent in the female not be activated either by an arid and sterile presence, and by seemingly barren and purely carnal movements of the male, or else by the violence and the repression of desires induced by the female; could they not then leave their storage site for the womb where they remain, and combine of their own accord? Could the mola not be the product of this combination of elements emanating from the female alone, or those originating just from the male? If the mola results from a combination such as I envisage, however, the laws governing this combination will be just as invariable as the laws of generation itself... The organisation of the mola will therefore remain invariable.[46] If we were to take up a scalpel and perform dissections on these molæ, we might even discover some which bore traces associated with the difference between the sexes. This might be described as the art of proceeding from the relatively unknown to the completely unknown. It is an irrational form of behaviour found to a surprising degree in those who have acquired, or who possess naturally, a gift for the experimental sciences; dreams of this sort have led to a number of discoveries. It is this sort of guesswork which should be taught to learners – if, indeed, it can be taught at all.

2. However, if it is discovered in the course of time that the mola is never engendered in the female without the involvement of the male, then a number of new and far more convincing conjectures can be formulated on the subject of this singular body. The web of blood-vessels which we call the placenta is known to be a mushroom-shaped segment of a sphere, adhering by its convex area to the womb throughout pregnancy, with the umbilical cord serving as a stalk; it comes away from the womb during the contractions of childbirth; its surface is smooth in a healthy woman who has had a successful delivery. In its generation, its bodily conformation and its behaviour, a living being is never anything but what the constraints of its existence, the laws of motion, and the universal order of things determine that it shall be; should this segment of a sphere – which appears to adhere to the womb only through being placed in contact with it – happen gradually to come away at the edges from the start of pregnancy, and continue to do so at a rate directly proportional to its increase in volume, it occurred to me that these edges, once free of any attachment, would draw closer and closer together, and assume a spherical shape. The umbilical cord, tugged by two opposing forces – one caused by the convex and detached edges of the segment, which would tend to shorten it, and the other by the weight of the foetus, which would tend to lengthen it – would then be much shorter than under normal conditions. A time would come when these edges would meet and knit together, forming a type of ovum, in the centre of which a foetus, as abnormal in its organisation as in its production, would be found, obstructed, constricted and suffocated. This ovum would feed until the small surface area still connecting it came away entirely under its weight, and it fell into the womb unattached, whence it would be ejected by being laid, rather as a hen lays an egg (an object to which the ovum, at least by its shape, has some similarity). If these conjectures could be tested in a mola, and if it were nonetheless demonstrated that this mola is engendered in the female without any assistance from the male, it would then clearly follow that the foetus is formed entirely in the female and that the male only becomes involved at the development stage[47].

XXXIII. *Conjectures: second series*

Let us assume that, as one of our greatest thinkers claims,[48] the earth has a solid vitreous core, and that this core is coated with dust; we may be sure that, in accordance with the laws of centrifugal force (which tend to push free bodies towards the equator, and to give the earth the shape of a flattened sphere), this dust would settle in finer layers at the two poles than at any other latitude; the core may be devoid of dust at the two ends of the axis, and to this characteristic should be attributed the direction in which a magnetised needle turns, no less than the aurora borealis (which is probably no more than an electrical current).[49]

There are good reasons for believing that magnetism and electricity arise from the same causes.[50] Why should they not be effects of the earth's rotational movement, and of the energy given off by the matter of which the earth is composed, combined with the action of the moon? High and low tides, currents, winds, light, and the movement of the free particles of the earth – and possibly even the motions of the entire earth's crust over the core, and so on – produce constant friction in innumerable ways. Over the centuries, the effects of observable and continuous causes achieve a great deal. The earth's core is a vitreous mass; its surface is covered merely with glass débris, sand and vitreous matter. Glass is, of all substances, the one which produces the greatest quantity of electricity by friction. Could the total mass of all the electricity on earth not be the result of all the friction taking place either at the surface of the earth or at its core? But it may be assumed from this general cause that a few attempts will be enough to deduce a particular cause which will create between two great phenomena – I refer to the position of the aurora borealis and the direction of a magnetised needle – a link similar to the connection observed between magnetism and electricity; this could be done by magnetising needles by means of electricity alone, without using a magnet. These notions can be either accepted or denied, since they only exist in my mind as yet. It is experimentation which will validate them, and the task of the empirical scientist is to devise experiments which will finally either separate or bring together different phenomena.[51]

XXXIV. *Conjectures: third series*

Electrical matter diffuses an observable sulphurous odour at sites where electricity is generated; did chemists not have the right to investigate this property? Why did they not test fluids with the highest possible electrical charge, using all the means at their disposal?[52] As yet, we do not even know whether electrolysed water dissolves sugar more or less rapidly than ordinary water. The heat of a furnace is known to produce a considerable increase in the weight of certain materials such as calcinated lead;[53] if electrical heat constantly applied to this calcifying metal were to increase this effect further still, could a new analogy then not be drawn between electrical heat and ordinary heat? This extraordinary type of heat has been tested to see whether it could enhance existing remedies, and make a substance more effective, or a topically-applied treatment more active; but were these trials not abandoned too soon? Why should electricity not modify the formation and properties of crystals[54]? Any number of conjectures are waiting to be formed in the mind - and to be confirmed or refuted by experiment! *See the next paragraph.*

XXXV. *Conjectures: fourth series*

Do most meteorites, will-o'- the-wisps, exhalations, shooting stars, natural or synthesised phosphorus, and rotting, luminous wood have any causes other than electricity?[55] Why should the experiments needed to determine the question not be conducted on these types of phosphorus? Why is there a reluctance to establish whether air, like glass, is an electrical body in itself - that is, a body which needs nothing more than to be rubbed and struck in order to be electro-lysed?[56] Who can tell whether air charged with sulphurous matter may be found to be more or less electrically charged than fresh air? If a metal rod offering a wide surface area is swung round at high speed in the air, it will be possible to determine whether the air is electrically charged, and how much electricity the rod has received from it. If, in the course of the experiment, sulphur is burnt along with other materials, those which increase, and those which reduce, the electrical quality of the air will then be identified. The cold air at the two poles may be more susceptible to electricity than the warm air of the equator; and, since ice is electrical and water is not,[57] who can say whether the enormous amount of permanent ice which is amassed at the poles - and which perhaps moves across the vitreous core, which is more exposed at the poles than elsewhere - might not account for the phenomena of the magnetised needle and the onset of the aurora borealis, both of which also appear to depend on electricity, as we suggested in the second series of these conjectures?[58] By observation, one of the most generalised and powerful forces of nature has been identified; it is now the task of experimentation to discover its effects.

XXXVI. *Conjectures: fifth series*

1. If the string of a musical instrument is taut, and an obstacle which weighs very little divides it into two unequal sections in such a way that the vibrations from one of these sections can be communicated to the other without any hindrance, we know that this obstacle causes the greater of the two sections to divide into vibrating segments, so that the two sections of the string sound in unison, and the vibrating segments of the larger section each fall between two fixed

points.[59] The resonance obtained in this way does not cause the division of the larger section, while the unison between the two sections is merely an effect of this division. It therefore occurred to me that, if a metal rod were to be substituted for the string of the instrument, and struck with great force, bulges and nodes would then form all along it; the same would be true of any elastic body, whether capable of producing sounds or not; this phenomenon, which is thought to be peculiar to vibrating strings, occurs to a greater or lesser extent whenever there is any percussion, and obeys the general laws governing the transmission of movement. Furthermore, in bodies which suffer impact, there are infinitely tiny oscillating segments, and stationary nodes or points which are immeasurably close together; these oscillating segments and these nodes are the cause of the tremor which we can feel upon touching the body after the impact, sometimes in the absence of any local transmission and sometimes once it has ended. This assumption fits in with the nature of the tremor, which does not run from the whole of the surface which has been touched to the whole of the surface of the sensitive area which touches, but from innumerable points (which are distributed over the surface of the body which has been touched) vibrating confusedly between infinite numbers of stationary points; apparently the force of inertia in continuous elastic bodies, evenly distributed throughout the mass, serves at a given point as a small obstacle in relation to some other point. Furthermore, assuming that the struck section of a vibrating string is infinitely small – and hence has infinitely small bulges, and nodes occurring immeasurably close together, an impression can be gained from this linear (and, so to speak, unidirectional) event of what takes place multi-directionally in one solid body impacted by another. As the length of the intercepted section of the vibrating string is a given, there is nothing which can cause the number of stationary points to be multiplied in the other section; since the number of these points is the same, irrespective of the force of the impact,[60] and since it is only the speed of the oscillations which varies, then, in the impact between the bodies, the tremor will have greater or lesser force; yet the relationship between the number of vibrating and stationary points will be the same, and the amount of matter in repose in these bodies will remain constant, whatever the force of the impact, the density of the body and the cohesion between the parts.[61] All the geometrician would then have to do would be to extend the calculations for the vibrating string to prisms, spheres and cylinders in order to establish the universal law of the distribution of motion in an impacted body - a law which, until now, we had not even begun to look for, since the existence of this phenomenon had not even been conceived, and we had supposed quite the reverse: motion had been assumed to be evenly distributed throughout a mass, even though, upon impact, the tremor indicated to the senses the presence of vibrating points distributed amongst stationary ones. I use the words 'upon impact,' since it is likely that in the transmission of motion in which impact plays no part, a body is projected in the same way as the smallest molecule would be, and that movement is uniform throughout the entire mass simultaneously. The tremor is non-existent in such cases, and this serves to distinguish instances of impact from those in which it does not occur.

2. All the forces acting on a given body can always be reduced to only one by using the principle of the resolution of forces; if the degree and direction of the force acting on the body are given, and an attempt is made to determine the resulting movement, it will be found that the body proceeds as if the force were passing through the centre of gravity and that, in addition, it rotates around the centre of gravity, as if the centre were fixed and the force were acting around the centre as around a fulcrum. Hence, if two molecules are attracted to each other, they will arrange themselves in relation to each other according to the laws governing their attraction, configuration, etc. If this system consisting of two molecules attracts a third, and is in turn attracted by this third molecule, these three will arrange themselves in relation to one another according to the laws governing their attraction, configuration, etc., and the same will be true of other systems and other molecules. They will form an entire system (A) in which, irrespective of whether or not they touch and whether or not they move, they will be resisting a force which would tend to disrupt the co-ordination between them; they will continually tend either to return to their initial order (if the disruptive force should happen to desist), or else to co-ordinate their actions in relation both to the laws governing their attraction, configuration, etc. and to the action of the disruptive force (if it persists). This system (A) is what I shall refer to as an 'elastic' body.[62] In this general, abstract sense, the solar system and the universe constitute an 'elastic' body: chaos cannot exist, for there is an order which necessarily ensues from the primary properties of matter.[63]

3. If we consider system (A) in a vacuum, it will be indestructible, immutable and eternal; if we suppose its constituent parts to be

disseminated in the vast expanse of space, just as properties such as attraction extend *ad infinitum* once there is nothing to restrain their field of action,[64] then these parts, whose configurations will not have varied at all, and which will be subject to the action of the same forces, will again co-ordinate as before, and will once more form, at some point in time and space, an elastic body.

4. This would not be the case if we were to take system (A) as existing in the universe; the ensuing effects would be no less necessary, although sometimes a fully-determined causal action could not possibly be at work; the causes which combine are always so numerous in the overall system or the universal elastic body, that we know neither what individual systems or elastic bodies may have been like originally, nor what they will become. Thus, without claiming that, in a plenum, attraction constitutes solidity and elasticity as we know them, is it not obvious that, in a vacuum, this property of matter is sufficient on its own to constitute these qualities, and to give rise to rarefaction, condensation and all the other phenomena which depend on them? Why, then, could it not be the primary cause of these phenomena in our overall system, in which an infinite number of causes working to modify this same property could bring about infinite variations in the numbers of these phenomena within particular systems or elastic bodies? Hence, an elastic body, when stretched, will break only when the cause drawing its parts together in one direction has drawn them so far in the other that they will no longer exert any noticeable attraction on one another; an elastic body will only burst upon impact when several of its vibrating molecules have been drawn, during their initial oscillation, such a distance from the stationary molecules amongst which they are distributed that they no longer exert any noticeable mutual attraction upon one another. Should the force of the impact be so great that the vibrating molecules were all drawn beyond their effective sphere of attraction, the body would be reduced to its elements. But, between such an impact – the strongest a body can experience – and one which would only produce the faintest tremor, there exists a degree of impact (which may be real or simply one which we can imagine), following which all constituent elements of the body, having separated, would no longer touch one another without their system being destroyed and their co-ordination ceasing. We shall leave to the reader the task of applying these same principles to condensation, rarefaction, etc. It only remains for us to reiterate the difference between the transmission of motion through

impact, and in the absence of impact. All the parts of a body, in the absence of impact, are transported equally and simultaneously; whatever the degree of motion imparted by this means, even if it were infinite, the body will not be destroyed, but will remain intact until some impact causes some of its components to oscillate amongst others which remain stationary; the peak of the first oscillations then reaches an amplitude at which the oscillating parts can no longer either return to their starting-point or become systematically co-ordinated once more.[65]

5. All the foregoing observations apply, strictly speaking, only to simple elastic bodies or systems of particles of the same matter and with the same configuration, activated by the same degree of force and driven by the same law of attraction. But if all these properties are variable, this will give rise to an infinite number of heterogeneous elastic bodies. By this I mean a system made up of two or more systems of different types of matter, with different configurations, activated to varying degrees, and perhaps even moving in accordance with, different laws of attraction, with particles co-ordinated among themselves by a law governing them all, which may be regarded as the product of their interaction.[66] If this composite system could be simplified by a few scientific operations to expel all the particles from one type of co-ordinated matter, or if it could be further compounded by introducing a new type of matter with particles interacting with those of the system, and hence changing the law governing all of them; then, the solidity, elasticity, compressibility, potential for rarefaction and other factors dependent – in the composite system – on the differing degrees of co-ordination between particles, will then increase or diminish, and so on. Accordingly, lead, which is not particularly solid or elastic, declines further in solidity and increases in elasticity when smelted, a process which consists of making a system composed of lead molecules interact with another consisting of air molecules, fire molecules, etc. which together make it into molten lead.[67]

6. These ideas could very easily be applied to innumerable other similar phenomena, and a lengthy treatise could be written on them. The hardest discovery to make would be to identify the mechanism whereby parts of a system, when they co-ordinate with the parts of another system, sometimes simplify the latter, driving out from it a system of other co-ordinated parts, as happens in certain chemical procedures. Attraction governed by different laws does not appear to offer an adequate explanation of this phenomenon, and it is hard to

admit the possibility of repulsive properties. This difficulty could be avoided as follows: Let us postulate a system (A) made up of systems (B) and (C), the molecules of which co-ordinate with one another according to some law applying to all of them. If we introduce a fourth system (D) into composite system (A), one of two things will happen: either the particles belonging to system (D) will co-ordinate with the constituent parts of system (A) without involving any impact, in which case system A would be made up of systems B, C and D; or, alternatively, co-ordination between particles belonging to system (D) and those of system (A) will be accompanied by an impact. If the impact is such that the particles which suffer it are not carried by their initial oscillation beyond the infinitely small sphere of their attraction, then disruption or innumerable slight oscillations will occur initially. But this disruption will soon end; the particles will co-ordinate, and from their co-ordination will result a system (A) made up of systems (B), (C) and (D). Should the constituent parts of system (B), those of system (C), or of both, be impacted at the very onset of co-ordination, and carried beyond their sphere of attraction by components of system (D), they will then be separated from the process of systematic co-ordination, and will not return to it again; system (A) will thus be a system composed either of systems (B) and (D) or of systems (C) and (D), or else it will be a simplified system consisting solely of the co-ordinated particles of system (D). These phenomena will occur in circumstances which may either greatly enhance – *or totally demolish* – the probability of these ideas. It should be added that I have reached this stage by taking as my starting-point *the vibration of an elastic body under impact.* Separation will never be spontaneous in the presence of *co-ordination*, although it could take place spontaneously where there is *composition* only. *Co-ordination* is also a principle associated with *uniformity,* even in a heterogeneous *whole.*[68]

XXXVII. *Conjectures: sixth series*

What the arts produce will remain commonplace, imperfect and feeble so long as no attempt is made to imitate nature more rigorously. Nature is slow and stubborn in its workings. It always advances towards its goals by the most imperceptible stages, whether distancing, approaching, combining, dividing, softening, condensing, hardening, liquefying, dissolving or assimilating. Art, on the other hand, makes haste, tires and slackens. Nature takes centuries to form metals crudely, while art sets out to perfect them in a single day. Nature takes centuries to shape precious stones, whereas art aims to counterfeit them in an instant. Even if we had the right means at our disposal, it would not be enough; we would still need to know how to apply them. It would be a mistake to imagine that, as the product of the intensity of the action multiplied by the time it occupies is the same, the result will also be the same. Transformation can be achieved only through slow, gradual and continuous application. Any other kind of application is merely destructive. How much we would gain by combining certain substances which yield only very imperfect compounds, if we were to proceed in a similar way to nature! But we are always impatient to have the satisfaction of completing anything we have begun. This explains all our fruitless endeavours, all our wasted expenditure and effort; so many works are suggested by nature which art will never attempt to emulate, because the chances of success seem so remote. Who was it who came out of the caves at Arcy unconvinced, from the speed at which stalactites are formed and grow again when they are damaged, that these caves will one day fill up and consist of nothing but a huge solid mass?[69] What naturalist, having pondered this phenomenon, has not postulated that by causing water to filter gradually through the ground and rock, and to drip into great empty caverns, it might eventually be possible to form quarries of artificial alabaster, marble and other types of precious stones, with properties which would vary depending on the nature of the ground, the water and the rock? But what is the use of these views without the stamina, patience, toil, expenditure, time and, above all, that taste which the Ancients had for great projects, to which so many memorials still exist, but to which we react with only cold and sterile admiration?

XXXVIII. *Conjectures: seventh series*

So many unsuccessful attempts have been made to convert our iron into steel equal to that of England and Germany, so that it could be used in the manufacture of delicately-crafted products.[70] I do not know what processes have been followed; it seems to me, however, that we would have been led to make a major discovery by emulating and perfecting an operation widely used by workers in iron-foundries. It is known as *puddling*. Puddling consists of taking the hardest soot, grinding it down, diluting it with urine, adding crushed garlic, old shoe leather which has been shredded, and common salt. Taking up a metal container, cover the base with a layer of this mixture, and place on it another layer of various pieces of ironwork, followed by a layer of the mixture, and so on, until the container is full. Replace its cover and coat the outside neatly with a mixture of heavy, compressed earth, wadding and horse dung. Place the container at the centre of a heap of charcoal proportionate to its volume; light the charcoal, leave the fire to burn, and just make sure that it does not go out. Take a container filled with cool water and, three or four hours after setting the container on the fire, remove and open it, dropping the pieces inside it into the cool water, stirring as the pieces are dropped in. These pieces are tempered by puddling and if some of them are broken apart, it will be found that the surface is covered with a very hard, fine-grained layer of steel to a shallow depth. This surface takes on a brighter polish, and will retain more effectively the shape it is filed into. Is there not a good chance that, once it has been exposed, layer upon layer, to the action of the fire and of the materials used in puddling, carefully-selected iron, which has been properly worked and made as thin as sheet-metal, or into very fine rods, and thrust, the moment it leaves the steel-making furnace, into a stream of water suitable for this process, would then be converted into steel? Especially if the task of conducting these initial experiments were entrusted to men who, long accustomed to dealing with iron, able to recognise its properties and to remedy its defects, would be sure to simplify the procedures and find materials more suitable for the process.

XXXIX. Will the presentation of experimental science in public lectures be enough to inspire the sort of philosophical frenzy I have described? I very much doubt it. Lecturers who perform experiments rather put one in mind of the man who imagines that he has given people a large meal because he has a crowd at his table. The main aim (pursuing the analogy) should be to whet the appetite so that some people, carried away by the urge to satisfy it, progress from being followers to being enthusiasts, and ultimately to taking up the profession of philosopher.[71] Every public figure should distance himself from such reservations, which are utterly inimical to scientific progress. Things in themselves should be revealed, as well as the means of achieving them.[72] The first men who discovered modern calculus through their power of invention were, in my view, very great men indeed! And how petty they seem to me in the mystery they created around it! Had Newton hastened to speak out, as the interests of his own glory and of the truth demanded, Leibnitz would not now share with him the title of joint inventor.[73] The German thought of the instrument, whereas the Englishman delighted in amazing scholars with astonishing ways of applying it. In mathematics and in the natural sciences, the most reliable way to start is to establish one's proprietorial rights by presenting one's credentials to the public. I should add that when I urge that the methods used should be revealed, I refer to those which have led to success; those which have not should be dealt with as succinctly as possible.

XL. It is not enough to reveal something; any revelation should be clear and complete. There is a type of obfuscation which could be defined as *the affectation of the great masters*. It serves as a veil which they enjoy drawing between nature and the public. Setting aside the respect due to famous names, I take the view that this kind of obscurity prevails in some of Stahl's* works and in Newton's *Principles of Mathematics*.[74] These books needed only to be understood to be valued as they should be, and their authors would have needed only a month to make them clear – a month which would have spared a thousand good minds three years of labour and fatigue. This comes to about three thousand wasted years in all, which could have been spent on other things. Let us hasten to popularise philosophy. If we want philosophers to lead the way, let us see that the public comes close to the point which the philosophers have reached. Will they say that there are works which can never be made accessible to ordinary minds? If they do, they will only be showing that they do not know what the right approach and force of habit can achieve.

If certain writers were entitled to be obscure (and at the risk of being accused of composing my own apologia here),[75] I venture to say that this would apply only to metaphysicians. Great abstractions glow with only a pale light. The very act of generalising tends to strip concepts of anything tangible, and as it progresses, so physical spectra fade away; concepts gradually move away from the realm of imagination towards that of the understanding, and ideas become purely intellectual. The speculative thinker then comes to resemble a man who watches from those mountain tops whose summits are lost in the clouds: the objects visible on the plain have disappeared before his eyes, so that all that remains with him is the spectacle of his own thoughts, and an awareness of the heights he has reached, where others may be unable to follow him and breathe the same air.

* The *Specimen Becherianum*; the *Zimotechnie*; the *Trecenta*; See the article 'Chimie' in vol. IV of the *Encyclopédie* (*Note by Diderot*).

XLI. Is nature not veiled enough already, without shrouding it in yet more mystery? Is our art not already beset with enough difficulties? Open a book by Franklin;[76] thumb through those written by chemists, and you will see how heavily the art of experimentation depends on opinions, imagination, sagacity and resources: read them through attentively because, if it is possible to discover in how many different directions an experiment can go, this is the place to find out. If, for lack of inspiration, you require some technical device to guide you, keep in front of you a table of the properties already observed in matter;[77] see which of these properties may be suited to the substance on which you intend to perform trials; make sure that they are there, and then attempt to measure them; this almost always requires an instrument, to which a constituent analogous to the substance can be applied uniformly and continuously, until nothing is left over and the property is completely exhausted.[78] The question of whether a property exists will only be resolved by means which are not immediately obvious. Even if we do not learn how to conduct the search, it is at least something to know what we are looking for. What is more, those who are forced to acknowledge inwardly their own barrenness, either because of their well-attested inability to make any discoveries themselves, or because they secretly envy the discoveries made by others, or because of the chagrin they feel despite themselves, and the petty manoeuvrings they willingly indulge in to share in the honour of a discovery, would do well to give up a science they practice without enhancing it – or adding to their own reputation.

XLII. Once one of those systems requiring experimental verification has taken shape in the mind, it is a mistake to adhere to it too stubbornly or to abandon it too readily.[79] We sometimes regard our conjectures as false without having taken the appropriate steps to establish that they are true. Here, stubbornness is actually less of a disadvantage than the opposite extreme.[80] By means of repeated trials, even if we fail to find what we are looking for, we may happen to find something better. Time spent consulting nature is never completely wasted. Its consistency should be measured by force of analogy.[81] Completely outlandish ideas do not warrant more than one trial. Those which look more plausible should be pursued, and those which appear to be leading to a significant discovery should not be abandoned until one feels defeated. I do not think there is any need to labour the point. Our dedication to research is naturally proportionate to the interest we take in it.

XLIII. As the systems under discussion are propped up by nothing more than vague ideas, mild suspicions, deceptive analogies - and yes, to be frank – by illusions which the excited mind readily takes to be valid viewpoints, no system should be dropped without first being put to the 'inversion' test.[82] In exclusively rationalist philosophy, truth is quite often taken as the complete opposite of error; similarly, in the experimental school, the phenomenon which was expected may be produced not by the experiment which has been carried out, but by its contrary. It is important to given special consideration to the two diametrically opposed points. This is why (to return to the second series of these reveries),[83] having covered the equator of the electrical globe and left the two poles uncovered, the poles should then be covered and the equator left uncovered; and, since it is important to make the simulation resemble the real globe as closely as possible, the choice of material used to cover the poles will also have some bearing on the question. It might be necessary to introduce accumulations of a fluid, as it is perfectly possible to do in practice, making the experiment yield some remarkable new finding which may differ from what was originally envisaged.

XLIV. Experiments must be repeated to establish the relevant conditions in detail, as well as to determine limits. They should be applied to different objects, and made more complex, in all their permutations and combinations. Whenever experiments have been sketchy, isolated, unconnected and incapable of being reduced to a simple result, this last characteristic itself shows that there is still a good deal to be done. We must then take hold of our subject and worry at it, so to speak, until the phenomena have been so firmly linked together that, if one is given, all the rest will follow. Let us first set about reducing effects to a simpler form; the reduction of causes will come afterwards. Effects can never be reduced to simple terms except by increasing their numbers. In employing the means to extract as much as possible from a given cause, the great art is to distinguish properly between those which might be expected to yield a new phenomenon and those which are capable only of producing a known phenomenon in a different disguise. Dwelling endlessly on such metamorphoses leads to exhaustion and stifles progress. Any experiment which does not extend the application of a law to some new case, or which does not use some exception to qualify it, is meaningless. The shortest way of discovering the value of a trial one is making is to take it as the first premise of an enthymeme[84], and then examine what follows. Is the consequence exactly the same as the findings from another trial? If so, no discovery will have been made; at most, a discovery will have been confirmed. There are few weighty tomes on experimental science which cannot be reduced to a few pages by this extremely simple rule; and plenty of smaller works might be reduced to nothing at all.

XLV. Just as, in mathematics, we find, on examining all the properties of a curve, that the same property is present throughout in different guises, so it will also be recognised, when experimental science is more advanced than at present, that all phenomena, whether we refer to weight, elasticity, attraction, magnetism or electricity, are no more than different aspects of the same property.[85] But, amongst the known phenomena which are attributed to one of these causes, how many intermediate phenomena are still waiting to be discovered before connections can be created, voids can be filled, and identity can be demonstrated? This we cannot determine. There could be some central phenomenon at work which sheds its light not just on those phenomena which are already known, but also on all those which will emerge in time, uniting them and arranging them into a system. But in the absence of this unifying focal point, they will remain in isolation; all the discoveries of experimental science will serve only to draw them closer together, without ever uniting them; and, should such discoveries ever succeed in uniting them, the result would be a continuous circle of phenomena, in which the first could not be distinguished from the last. This singular occurrence, whereby experimental science, by sheer effort, would have produced a maze in which rationalist science could only wander round and round, helpless and lost, is not impossible in the natural sciences as it is in mathematics. In mathematics, intermediate propositions can always be found, by synthesis or analysis, which separate the fundamental property of a curve from the most remote of its properties.[86]

XLVI. There are deceptive phenomena which, at first sight, appear to overturn a system but which, if they were better known, would finally substantiate it.[87] Such phenomena become a torment for the philosopher, especially when he has the feeling that nature is inflicting it on him, and shielding itself from his speculations by some strange and secret mechanism. This uncomfortable situation occurs whenever a given phenomenon arises from several conjoined or conflicting forces. When they are conjoined, the scale of the phenomenon will be too great to fit the hypothesis which has been formed; if they conflict, the scale will be too small. It may even shrink to nothing and the phenomenon will disappear, without our having any idea how to explain this sudden, unpredictable silence of nature. Do we begin to suspect the reason why this happens? That really doesn't get us much further. An effort must be made to separate the causes, to analyse the results of their actions, and to reduce a very complicated phenomenon to a simple one; or at least to reveal, by some new experiment, the complex nature of these causes, and the ways in which they complement or oppose one another; this is often a delicate, and sometimes impossible, operation. Then the system totters; philosophers are divided in their opinions; some still adhere to the system, while others are swept along by the experiment which appears to refute it. The disagreement continues until either wit or chance (which is never idle - and is more fruitful than man's wit) removes the contradiction and reinstates ideas which had been all but abandoned.[88]

XLVII. Experimentation must be given a free rein; to show only what confirms our findings and conceal what conflicts with them would be to hold it captive. Therein lies the drawback, not in having ideas as such, but in letting oneself be blinded by them when carrying out an experiment.[89] We use our critical judgement only when the result conflicts with the system. Nothing is then omitted which could alter the aspect of the phenomenon, or the language of nature. If the opposite occurs, the observer is indulgent; he glosses over the circumstances, and hardly bothers to raise objections in the face of nature; he takes her at her word; he suspects no ambiguity, and he deserves to be told 'Your task is to ask questions of nature, but you are either making her lie, or else you are afraid to make her explain herself.'

XLVIII. When we have taken the wrong road, the faster we walk, the more we goes astray. But how can anyone retrace his steps once he has covered a huge distance? Sheer exhaustion does not allow him to do so; and vanity, too, bars our way without our realising it; a stubborn attachment to principles gives everything surrounding us an illusory appearance which distorts objects. Things are then no longer seen as they are, but as they ought to be. Instead of modelling our ideas on real beings, it seems that we make every effort to construe beings on the basis of our ideas. Of all the different schools of thought, none is so clearly overcome by this craze as the 'methodists'.[90] As soon as one of their number has devised a system which places man at the head of the quadrupeds, he sees him in nature as nothing more than a four-footed animal. The incomparable reasoning faculty with which man has been gifted cries out against the term 'animal', and his organism belies the description of 'quadruped', but in vain; nature may well make man look heavenwards - his systematising bias will still bend his body towards the ground. This same bias leads people to maintain that reason is merely a more perfect form of instinct, and to believe, in all seriousness, that it is only because he has grown out of the habit that man loses the use of his legs when he decides to transform his hands into two feet.[91]

XLIX. Indeed, the arguments used by some of the classifiers are so strange that an example needs to be given here. Man, says Linnæus in the preface to his *Fauna Suecica*,[92] is neither a stone nor a plant; he must therefore be an animal. He does not have only one foot, so he cannot be a worm. He is not an insect because he has no antennae, nor a fish, because he has no fins, nor a bird, because he has no feathers. So what is man? He has a mouth like a quadruped. He has four feet; he uses the two fore-feet to touch with, and the two hind-feet to walk with. So he must be a quadruped. The Linnæan goes on to say: 'It is true that, as a consequence of my theories about natural history, I have never been able to distinguish between man and ape, because there are certain apes which have less hair than certain men: these apes walk on two legs and use their hands and feet like men. Nor do I consider speech to be a distinguishing feature; my method only allows for those features which are dependent on number, contour, proportion and situation.' 'So your logic must be wrong', the logician will say; and the naturalist will conclude that man is a four-footed animal.[93]

L. To shatter a hypothesis, it sometimes needs only to be taken to its logical conclusion. We shall now put this approach to the test in the case of the doctor of Erlangen,[94] whose book, teeming as it is with extraordinary new ideas, will put any philosopher on the rack. His subject is the greatest that human intelligence can address: nature as a universal system. The author begins with a rapid exposé of the views of his predecessors, and the reasons why their principles are inadequate to explain the overall development of phenomena. Some philosophers require only *extension* and *motion*.[95] Others have found it necessary to enlarge upon extension by adding *impenetrability, motility* and *inertia*.[96] Observation of the heavenly bodies, or the scientific study of large bodies in general, has pointed to the need to postulate a force or law according to which all the parts are drawn towards, or weigh upon, one another; and the force of *attraction*, in direct ratio to mass and in inverse proportion to the square of the distance, is now generally accepted.[97] The simplest chemical procedures, or the elementary physics of small bodies, have entailed a resort to forms of *attraction* which follow other laws; the impossibility of explaining the formation of a plant or animal by means of attraction, inertia, motility, impenetrability, motion, matter or extension has led the philosopher Baumann to attribute yet further properties to nature. Dissatisfied with the idea of a *plastic nature*, which can be made to perform all the wonders of nature without resorting to matter or intelligence;[98] rejecting *subordinate intelligent substances*, which exert some unintelligible action on matter; refusing to accept the *simultaneity of creation and the formation of substances* which, being contained within one another, develop over time as a continuation of the first miracle;[99] and, lastly, rebutting the idea that *they are produced extemporaneously*, which makes them nothing more than a chain of miracles repeated with each moment that passes, he has come to the view that all these rather unphilosophical systems would never have arisen were it not for a groundless fear of attributing well-known modifications to a being whose essence is unknown to us - and who is, for this very reason, despite our preconceptions, entirely compatible with such modifications. But what is this being? and what are these modifications? Dr Baumann asks: 'Shall I tell you?', and he answers: 'This being is a physical one; these modifications are *desire, aversion, memory* and *intelligence*' – in a word, all the qualities which we recognise in animals, which the Ancients referred to as a *sensitive soul*, and which Dr Baumann recognises, keeping form and mass in proportion, in anything from the smallest particle of matter to the largest of animals.[100] If, he says, there were any danger in attributing some degree of intelligence to molecules of matter, that danger would be equally great, whether we were dealing with an elephant or an ape, or even imagining it in a grain of sand. It is here that the philosopher of the Erlangen Academy uses every effort to deflect any suspicion of atheism on his part; and it is obvious that he hotly defends his hypothesis only because it appears to him to deal with the most challenging of phenomena, without leading inevitably to materialism.[101] Anyone who wishes to learn how to reconcile the boldest philosophical ideas with the deepest respect for religion should read his book. God created the world, says Dr Baumann, and it is up to us to find, if possible, the laws according to which He wished it to remain in being, and the means He has prescribed for the reproduction of the species. We are perfectly at liberty to put forward our ideas in this regard; and the doctor's main ideas are as follows:[102]

The seminal element issues from a part resembling the part which that element is later to form in the animal; it is sentient and thinking, and will retain some memory of its origins – this explains the preservation of species and the resemblance which offspring have to their forbears.

It may happen that seminal fluid has either an excess or a lack of certain elements and that, as a result of a lapse of memory, these elements cannot unite; it may also happen that strange combinations of superfluous elements are formed, resulting either in an inability to reproduce, or in the creation of every possible type of monstrosity.

Certain elements will inevitably have acquired a prodigious facility for always combining in the same way; it follows that infinitely varied microscopic creatures will be formed if the elements differ, and polyps if they are alike; the latter may be compared to a cluster of infinitely small bees which can store only one position in their memory, clinging to one another and remaining there in accordance with that position, which is the one most familiar to them.[103]

When the impression of the present situation counterbalances or extinguishes the memory of some past situation, resulting in indifference to all situations, sterility will ensue; that is why mules are sterile.

What is there to prevent primary intelligent and sentient components from diverging immeasurably from the pattern which determines a species? This would give an infinite number of animal species derived from one original animal; an infinite number of beings deriving from one original being: in short, a single act of nature.[104]

But, in the process of accumulation and combination, could each element lose its small store of feeling and perception?[105] Not at all, according to Dr Baumann. These qualities are vital to it. So what will happen? This: from these perceptions of assembled and combined elements, a single perception will arise, which is proportionate to the mass and to the configuration of what is perceived; and the system of perceptions in which each element has lost its recollection of the *self*, and will play its part in forming the consciousness of *the whole*, will be the creature's soul. *Omnes elementorum perceptiones conspirare, et in unam fortiorem et magis perfectam perceptionem coalescere videntur. Haec forte ad unamquamque ex aliis perceptionibus se habet in eadem ratione qua corpus organisatum ad elementum. Elementum quodvis, post suam cum aliis copulationem, cum suam perceptionem illarum perceptionibus confudit, et sui conscientiam perdidit, primi elementorum status memoria nulla superest, et nostra nobis origo omnino abdita manet.* *[106]

At this point, we are surprised that the author did not realise the dreadful consequences of his hypothesis; or, if in fact he did realise the consequences, that he did not abandon it.[107] The time has come to apply our method to testing his premises. Accordingly, I shall ask him whether the universe or the entire array of all sentient and thinking molecules forms a whole or not. Should he reply that it does not form a whole, then, with a single word, he will undermine the existence of God, by introducing disorder into nature, and he will destroy the very basis of philosophy by breaking the chain which links all beings together.[108] If he agrees that it does indeed form a whole, in which the elements themselves are no less well ordered than are their constituent parts (whether those parts actually differ from one another, or are merely thought of as doing so), and no less well ordered than these elements, in turn, are in any living creature, then he would have to allow that, as a result of this universal connection between things, the world has a soul, as though it were some great animal; and that, since the world may be infinite, this world-soul may be - although I do not say that it is - an infinite set of perceptions, and that the world could be God.[109] He can protest as much as he wishes against such consequences – they are nonetheless true; and whatever light his sublime ideas may cast upon nature's hidden depths, these ideas still remain just as terrifying. All it needed for this to become apparent was to apply them generally. Generalisation is, for a metaphysician's hypotheses, what repeated observation and experimentation are to the conjectures of the empirical scientist. Are these conjectures borne out? The more experiments are performed, the more these conjectures are verified. Are the hypotheses valid? The wider the consequences

---

* See paragraph 52, page 78 [of the *Dissertatio*]for this section; and in the pages which precede and follow it will be found some very astute and very convincing examples of how these principles can be applied to other phenomena (*Note by Diderot*).

range, the more truths they embrace, and the more compelling the evidence they provide. If, on the other hand, the conjectures and hypotheses are only weak and ill-founded, either some fact will be discovered, or some truth will emerge on which they founder. Dr Baumann's hypothesis may well unlock the most unfathomable mystery of nature – how animals or, more generally, organisms of any kind are formed – but he faces two pitfalls: the entire array of phenomena in the universe, and the existence of God. While rejecting the ideas put forward by the doctor of Erlangen, we would still have found difficulty in coping with the obscure phenomena which he set out to explain, the richness of his hypothesis, the surprising consequences which can be drawn from it, the value of his new speculations on a subject which has preoccupied the most outstanding figures throughout the centuries, and the difficulty of successfully challenging such speculations - had we not seen these things as the fruits of profound meditation, and a bold attempt by a great philosopher to tackle the entire system of nature.[110]

LI. *On the impetus of sensation*

If Dr Baumann had confined his system within reasonable limits, and had applied his ideas only to the formation of animals, without extending them to encompass the nature of soul (from which I believe I have shown, whatever he may say, that they could even be applied to the existence of God), he would never have rushed into the most seductive form of materialism by attributing desire, aversion, feeling and thought to organic molecules.[111] He should have been content to suppose that they were capable only of feelings a thousand times less intense than those which the Almighty has bestowed upon the stupidest creatures who are closest to lifeless matter. As a result of this subdued sensitivity and the difference in configuration, there would never have been more than one position to suit any given organic molecule, and that position would have been the most comfortable of all; this is the position which it would constantly have sought, with an unthinking restlessness, just as when animals stir in their sleep, and the use of almost all their faculties is suspended, until they find the position most conducive to their rest. This one principle would have accounted satisfactorily, simply, and without any dangerous consequences, for the phenomena which he set out to explain, and the innumerable wonders which so dazzle our entomologists. He would then have defined living creatures in general as a *system of different organic molecules which, under the impetus of a sensation similar to a dull, blunt feeling of touch granted to them by the creator of matter in its entirety, have combined with one another until each finds the place best suited to its configuration and its repose.*[112]

LII. *On instruments and measurements*

We have observed elsewhere that as the senses were the source of all our knowledge, it was most important to know how far we could rely on their testimony;[113] at this point it should be added that testing by using an extension of our senses – in other words, instruments – is equally essential. This is a new application of experimentation, and it means another source of long, arduous and difficult observations. There might be one way of cutting down the work: namely, to close one's ears to a kind of scruple attaching to rationalist philosophy (for it does have its scruples), and to have a proper grasp of whether the accurate measurement of quantities is necessary in all cases.[114] Think of all the effort, work and time wasted on measurement which could have been better spent on discovery!

LIII. In both the invention and the perfecting of instruments, there are precautions which cannot be too strongly recommended to the empirical scientist, i.e. to mistrust analogies;[115] never to extrapolate from major to minor instances, nor from minor to major ones,[116] and to scrutinise all the physical properties of the substances used. The empirical scientist will never succeed if he omits to do these things and, even when he has taken all the precautions he can, how often will it still happen that some minor obstacle, which he has either failed to foresee or which he has dismissed with contempt, will set a natural limit to his work, and force him to abandon it when he believed it to be complete?

LIV. *On distinguishing objects*

The mind cannot comprehend, nor the imagination predict, everything; the senses cannot observe, nor the memory retain, everything; great men are born at such rare intervals, and scientific progress is interrupted to so great an extent by revolutions, that centuries of study are required to recover the knowledge of past centuries; for these reasons, it would be failing the human race to provide imprecise observations of everything. Men of outstanding talent must respect both themselves and posterity in the way they use their time. What would posterity think of us, if all we had to leave behind was a complete entomology or a vast survey of microscopic creatures?[117] Let great minds tackle great subjects, and lesser minds concern themselves with lesser ones; they are better occupied with such matters than doing nothing at all.

LV. *On obstacles*

And since it is not enough to desire something, and since we have to accept at the same time everything which is almost inseparably linked to what is desired, anyone who has decided to apply himself to the study of philosophy should expect to encounter not just the physical obstacles associated with his subject, but also the numerous moral obstacles which will certainly beset him, as they have beset every philosopher before him. So, once he has been flouted, misunderstood, slandered, compromised and torn apart, he should be able to say to himself: 'Is it only in my century - and only in my case - that there have been people filled with ignorance and venom, spirits devoured by envy, and minds clouded by superstition?' If he sometimes believes that he has grounds to complain about his fellowmen, let him tell himself this: 'I do complain of my fellow-men, but, if it were possible to question all of them, and ask each one whether he would prefer to have been the writer of the *Nouvelles Ecclésiastiques* or Montesquieu, and the author of the *Lettres Américaines* or Buffon,[118] is there a single one who, if he had a little discernment, might have hesitated in his choice? I am therefore sure to receive some day the only plaudits I value, if I have had the good fortune to deserve them.'

And you, who take for yourselves the title of philosophers or intellectuals, and are not ashamed of resembling those troublesome insects which spend every moment of their fleeting existence disturbing man during his work and rest − what is your purpose? What do you expect from your furious efforts? Once you have disheartened the authors of any renown, and the outstanding geniuses which the nation still has, what will you then do in your turn for the national good?[119] What wondrous productions will you provide to compensate the human race for those it would have received? ... Despite what you do, the names of Duclos, d'Alembert and Rousseau, those of Voltaire, Maupertuis and Montesquieu, and those of Buffon and Daubenton, will still be honoured by us, and by our descendants; and if someone, some day, remembers your own names, he will say,

'They were the persecutors of the greatest men of their times; and if we now have the preface to the *Encyclopédie*, the *Histoire du Siècle de Louis XIV*, the *Esprit des Lois* and the *Histoire de la Nature*, it is because, fortunately, those people were powerless to deprive us of them.'[120]

LVI. *On Causes*.

1. Were we to consult only the vain conjectures of philosophy and the faint light of our reason, we should believe that the chain of causes had no beginning, and that the chain of effects will have no end.[121] Imagine that a molecule has moved: it has not moved of its own accord; the cause of its movement has another cause; this in turn has its cause – and so on, without our ever reaching the *natural* limits of causes in the past. Now imagine that a molecule moves; its movement will have its effect, and this effect in turn will have another effect, and so on, without our ever reaching the *natural* limits to effects in the future. The mind, terrified by this infinite progression towards the most trivial causes and the slightest effects, rejects this supposition - and certain others of the same type – only because it has a preconception that nothing takes place takes beyond the realms of the senses, and that everything beyond our view ceases to exist.[122] But one of the main differences between the *observer* and the *interpreter* of nature is that the latter begins at the point where the former ceases to use his senses and his instruments; on the basis of what now exists, he speculates on what is to come; from the established order of things, he draws abstract and generalised conclusions which, for him, have all the force of particular, ascertainable truths; he ascends to the very essence of the natural order, and sees that *the mere fact* that a thinking and sentient being co-exists alongside some kind of connection between causes and effects, is not in itself enough to enable him to make any definitive judgement on them; he stops at that point, and any further step he might take would put him beyond the boundaries of nature.[123]

2. *On final causes*. Who are we to explain nature's purposes? Do we not notice that, in commending her wisdom, we nearly always detract from her power, and that we are taking away more from her resources than we could ever attribute to her views?[124] This way of interpreting nature is wrong, even in natural theology,[125] as it substitutes human speculation for the workings of God, and binds the most important of truths to the fortunes of a hypothesis.[126] But the

most commonplace phenomenon will suffice to show how far the search for final causes is the opposite of true science. Supposing that an empirical scientist, when questioned as to the nature of milk, were to reply that it is a foodstuff which begins to form in the female after conception, and which nature intends for the nourishment of the future offspring – what would this definition teach me about how milk is formed? What am I to believe about the supposed purpose of this fluid, and the other physiological ideas accompanying it, when I know that there have been men who have made milk spurt from their breasts; I also know that anastomosis of the epigastric and mammary arteries* shows that milk is the cause of the swelling of the breasts which sometimes inconveniences young girls when their periods are due, and that nearly every girl could breast-feed if she were to suckle a child; in fact, I have before me a female who is so small in size that no male has been found to suit her; she has never mated and has never borne any young, but her nipples are nonetheless so engorged with milk that the usual means have had to be employed to bring her relief? How ridiculous anatomists appear when, in all seriousness, they attribute to coyness on the part of nature the shadow which she has also cast over other parts of the body where there is nothing indecent to hide! The purpose which other anatomists imagine that it serves does rather less honour to nature's modesty, but hardly any greater honour to their sagacity. The empirical scientist, whose profession is to instruct and not to edify, will therefore stop asking *why*, and concern himself only with the question of *how*. The question of *how* is based on actual beings, and the question of *why* is merely a product of our minds; it is associated with the systems we have invented, and depends on the progress of our knowledge. What a multitude of absurd ideas, false speculations and illusory notions are to be found in the hymns which a few headstrong proponents of final causes have ventured to compose in honour of the Creator! Instead of sharing the ecstatic admiration of the Prophet and calling out in the night, at the sight of the innumerable stars which brighten the heavens, *Coeli enarrant gloriam Dei*,[128] they have abandoned themselves to their superstitious conjectures. Instead of adoring the Almighty in nature's own creatures, they have bowed down before the figments of their imagination. Should anyone be bound by preconceptions and doubt

---

* This anatomical discovery, one of the finest to be made in our time, is the work of M. Bertin (*Note by Diderot*).[127]

the soundness of my reproach, I invite him to compare Galen's treatise on the function of the parts of the human body with the *Physiologie* of Boërhaave, and to compare the *Physiologie* of Boërhaave in turn with that of Haller.[129] I call upon posterity to compare whatever fleeting, systematic views are contained in the latter work with what physiology will become in future centuries. Man attributes the merits of his petty views to an eternal God; and eternal God, who listens to him from His lofty throne, and who realises his intentions, accepts his idiotic praises and smiles at his vanity.

LVII. *On various preconceptions*

There is nothing, either in the workings of nature, or in the conditions of life, which is not a pitfall for the hasty. As examples, I cite most of the popular axiomatic expressions regarded as 'common sense' by every nation.[130] There is a saying that *there is nothing new under the sun*,[131] and that is true for anyone who does not go beyond mere outward appearances. But what sort of maxim is it for the philosopher, whose daily task consists of grasping the most imperceptible of differences? How would it seem to someone who held that, on a whole tree, no two leaves would be *perceptibly* of the same shade of green?[132] How would it be regarded by someone who, speculating on the large number of causes (including those already known) which need to combine in order to produce a colour of exactly the same shade, claimed – without intending any offence to Leibnitz's view – that it has been shown, through the difference between the points in space at which bodies are placed, combined with the prodigious number of causes involved, that perhaps there never was, and perhaps never will be in the whole of nature, two blades of grass of *exactly* the same shade of green? If beings change in gradual stages by passing through the most imperceptible of nuances, then time, which does not stand still, must eventually create the greatest possible difference between forms which existed in the remote past, those existing today, and those which will exist in the distant future;[133] and so the saying *nil sub sole novum* is no more than a preconception based on the inadequacy of our organs, the imperfections of our instruments, and the brevity of our lives. There is a maxim of moral philosophy which says *tot capita, tot sensus* ('there are as many opinions as there are heads to hold them'), but in fact the opposite is true: nothing is so plentiful as heads, and nothing is so rare as opinions. There is a literary saying that *there is no arguing about taste*;

if this means that nobody should argue with a man that his taste is such and such, it is mere childishness. If it means that there is no such thing as good or bad taste, it is false.[134] A philosopher will subject all these expressions of popular wisdom to a stern examination.

LVIII. Questions[135]

Homogeneity can manifest itself in only one way, whereas there are an infinite number of different ways to express heterogeneity. It seems to me just as impossible for all nature's creatures to have been produced from one perfectly homogeneous type of matter as to represent them as all being of the same colour. I think there are even some indications that the diversity of phenomena could not have resulted from just any type of heterogeneity. Accordingly, I shall refer to the different types of heterogeneous matter needed for the general production of natural phenomena as *elements*, and I shall use the word *nature* to denote the existing overall result, or the successive overall results, of the combination of elements. There must be fundamental differences between elements, otherwise everything could have been produced by homogeneity, because everything could return to that state.[136] There either is, was or will be a natural or an artificial combination whereby one element is, was or will be divided as far as possible. The molecule of an element at this final point of division is indivisible; its indivisibility is absolute, since any further division of this molecule, being outside the laws of nature and beyond the power of our techniques, can only be theoretical.[137] The final point of division possible in nature would appear to differ from that lying within the power of artificial means, as far as fundamentally heterogeneous types of matter are concerned; it follows that there are molecules whose mass is essentially different, but which are absolutely indivisible in themselves. How many types of matter exist which are completely heterogeneous, or consist of one single element? We do not know. What are the essential differences between the types of matter which we regard as absolutely heterogeneous or composed of one single element? We do not know. How far can the division of matter composed of one element be taken, either by using man-made techniques or the workings of nature? We do not know - and so on, and so on. I have linked processes which require human intervention with those which occur naturally because, amongst the infinite number of facts which we do not know, and which we will never know, there is one which still remains hidden from us: namely, the question of whether an element has ever been, is now, or ever will be, split still further by some human technique than it has been, is now, or ever will be, split by any combination of forces when nature is left to its own devices. And it will be seen, from the first of the following questions, why I have introduced references to the past, present and future into some of my proposals, and why I have introduced the idea of succession into my definition of nature.[138]

1. If there is no link from one phenomenon to another, there can be no philosophy. Even if all phenomena were interlinked, the state of each of them might still not be permanent. But if each living being is in a perpetual state of change, even as a result of the workings of nature, then despite the chain which links phenomena together, there is still no philosophy.[139] All our natural sciences become as transitory as the words we utter. What we take for natural history is merely the far-from-complete history of a single instant. I ask, therefore, whether metals always have been, and always will be, as they now are; whether plants always have been, and always will be, as they now are; whether animals always have been, and always will be, as they now are; and so on. A word to sceptics: having meditated profoundly on certain phenomena, you may understandably question not so much the fact that the world was created, but whether it now is as it used to be, and as it will be in the future.

2. Just as in the animal and vegetable kingdoms, an individual comes into being, so to speak, grows, remains in being, declines and passes on, will it not be the same for entire species? If our faith did not teach us that animals left the Creator's hands just as they now appear and, if it were permitted to entertain the slightest doubt as to their beginning and their end, may not a philosopher, left to his own conjectures, suspect that, from time immemorial, animal life had its own constituent elements, scattered and intermingled with the general body of matter, and that it happened that these constituent elements came together because it was possible for them to do so; that the embryo formed from these elements went through innumerable arrangements and developments, successively acquiring movement, feeling, ideas, thought, reflection, consciousness, feelings, emotions, signs, gestures, sounds, articulate sounds, language, laws, arts and sciences; that millions of years passed between each of these developments, and there may be other developments or kinds of growth still to come of which we know nothing; that a stationary point either has been or will be reached; that the embryo either is, or will be, moving

away from this point through a process of everlasting decay, during which its faculties will leave it in the same way as they arrived; that it will disappear for ever from nature – or rather, that it will continue to exist there, but in a form and with faculties very different from those it displays at this present point in time?[140] Religion saves us from many deviations, and a good deal of work. Had religion not enlightened us on the origin of the world and the universal system of being, what a multitude of different hypotheses we would have been tempted to take as nature's secret! Since these hypotheses are all equally wrong, they would all have seemed almost equally plausible. The question of why anything exists is the most awkward that philosophy can raise – and Revelation alone provides an answer.[141]

3. If we glance at animals and the rough ground they tread, at organic molecules and the fluid in which they move, at microscopic insects and the matter which produces and surrounds them, it is obvious that matter is divided overall into two kinds: dead and living.[142] But how can it be that matter does not form a single unity, which is either wholly dead or wholly living? Is living matter always alive? And is dead matter really – and permanently – dead? Does living matter not die at all? Does dead matter never come to life?

4. Is there any ascertainable difference between dead and living matter other than its arrangement, and the real or apparent spontaneity of its movement?

5. Could so-called living matter not simply be matter which moves by itself? And could so-called dead matter not be one type of matter moved by another?[143]

6. If living matter is matter which moves by itself, how can it cease to move without dying ?

7. If there is matter which lives of its own accord, and matter which dies of its own accord, are these two principles sufficient to explain in general terms the production of all entities and all phenomena?

8. In geometry, a real parameter added to an imaginary one gives an imaginary whole; in nature, if a molecule of living matter is applied to a molecule of dead matter, will the whole then be living or dead?

9. If the combination can be either living or dead, at what point will it become living - and why? At what point - and why - will it be dead?

10. Whether living or dead, it exists in some form. In whatever form it exists, what principle governs it?

11. Are forms produced according to a matrix? What is a matrix? is it a real, pre-existing entity?[144] or does it merely designate the intelligible limits of a living molecule's energy, joined to dead or living matter, and determined by the relationship between energy in its widest sense, and resistance, also in the widest sense? If it is indeed a real, pre-existing entity, how was it formed?

12. Does the energy of a living molecule vary of its own accord? Or does it vary only according to the quantity, quality and shapes of the living or dead matter with which it unites?[145]

13. Are there any types of living matter identifiably different from other living matter? Or is all living matter basically of one type and suitable for everything? The same question can be asked with regard to dead matter.

14. Does living matter combine with living matter? How would it do so? What is the result? Again the same question can be asked with regard to dead matter.[146]

15. Supposing that all matter were living, or that all matter were dead - would there ever be anything other than dead matter (or living matter) or could living molecules not come back to life after losing their lives, only to lose them again, and so on, ad infinitum?

When I turn my gaze to the works of mankind, and I see towns built everywhere, materials of all kinds being used, languages established, nations policed, harbours constructed, seas crossed and the earth and skies measured; the world seems very old to me. But when I encounter people unsure as to the basic principles of medicine and agriculture, the properties of the commonest substances, the diseases affecting them, the height of trees and the shape of the plough, it seems to me that the earth has been peopled only since yesterday. And, if men were wise, they would finally devote themselves to research affecting their welfare, and would not answer my futile questions until at least a thousand years had passed; or perhaps, always bearing in mind how small a part of time and space they occupy, they would never even deign to reply.

THE END

Note to the reader concerning one point is section XXXVI, paragraph 3.

I mentioned, young man, that *properties such as attraction extend* ad infinitum *once there is nothing to restrain their field of action.* It will be objected 'that I could even have said that *they are propagated uniformly.* It will perhaps be added that it is scarcely possible to understand how a quality may be exercised at a distance, without any intermediary, but that there is not – and there never has been – anything absurd in this, although there would indeed be something absurd in claiming that it operates differently in a vacuum, at varying distances; for if that is so, there is nothing, whether inside or outside a portion of matter, which is capable of causing its action to vary; it may also be added that Descartes and Newton, as well as philosophers both ancient and modern, presumed that a body, powered with the slightest degree of motion in a vacuum, would continue *ad infinitum,* at a uniform rate and in a straight line; that distance in itself is thus neither an obstacle nor a vehicle; that any quality whose action varies in some proportion, whether direct or inverse, to distance, must necessarily lead back to the notion of the plenum, and to the corpuscular theory; and that the positing of a vacuum and of variability in causal action are two contradictory notions.' If anyone mentions these difficulties to you, I advise you to go and seek the answer from some Newtonian, for I confess that I do not know how to resolve them.[147]

Notes

1 This short prefatory address was added to the revised edition of 1754.

2 The basis of the charge of atheism levelled against Spinoza (1632–77) was that, in his *Ethica*[*Ethics*] (1677), he had made God and nature identical, and was therefore a materialist. La Mettrie (1709–51) had published his equally notorious *L'Homme Machine* [*Man a Machine*] in 1748, earning himself widespread condemnation as an even more intransigent materialist than Spinoza. Diderot's concern is to avoid being similarly vilified by the defenders of orthodoxy, who had already found much to decry in the 'philosophical' articles of the *Encyclopédie*. The reference to hypothesis is simply a reminder of its status in a work in which its role is central to the endeavours of the scientist.

3 'From the darkness, we behold those things which are bathed in light'. The quotation is taken from Book IV (not Book VI) of the *De rerum natura* [*On the Nature of Things*] (l.337). The reference is intended partly as a criticism of the obscurantism afflicting science at that time, and partly as a hint of the importance with Lucretian atomism will play in the *Interpretation*.

4 An echo of the preface to Bacon's *Novum Organum*.

5 The first volume of Buffon's *Natural History* had appeared in 1749.

6 Diderot is echoing Bacon who, in the *Advancement*, had written that '*Mathematick* and *Logick*, which should carry themselves as handmaids to *Physick*, boasting their certainty above it, take upon them a command and Dominion.' (Book III, Chap. VI).

7 'Metaphysics' in the sense that mathematics has nothing to do with the real world of *physis*, or nature.

8 'Rational' in the sense of 'expressible by numbers'.

9 James Bradley (1692–1762) was the first to measure accurately the diameter of Venus, and in 1747 had discovered the nutation, or slight oscillation, of the Earth's axis. Pierre-Charles Le Monnier (1715–99) had published his *Observations* on the sun, moon and fixed stars in 1751.

10 A reference to d'Alembert in particular, whose *Mélanges de littérature et de philosophie* (containing, *inter alia*, extracts translated from Tacitus) had been published in 1753.

11 This list contains the names of the most illustrious mathematicians of the day: Daniel Bernoulli (1700–82) and his brother Jean (1710–90). Leonard Euler (1701–83); Pierre-Louis Moreau de Maupertuis (1698–1759); Alexis Clairaut(1713–65); Alexis Fontaine des Bertins (1705?–71); Jean Le Rond d'Alembert(1717–83).

12 The Pillars of Hercules were two promontories (one in Spain, the other in Africa) at the entrance to the Mediterranean; the Ancients believed that they had been created by Hercules tearing the land apart, and that they marked the limits of the world accessible by man.

13 There may be a covert reference here to the very popular courses on physics given by the abbé Nollet (1700–1770) and others. But the last sentence indicates that Diderot is thinking particularly of the vogue created by the great success of Buffon's *Natural History*.

14 The tone of this whole section recalls the remarks of Christian apologists such as Blaise Pascal (1623–62) who had sought to humble man's vanity and pride in his achievements. The emphasis is particularly apt, since Pascal had been one of the great figures of mathematics in the previous century. On the importance of this

moralistic tradition in pre-Revolutionary France, see Anthony McKenna *De Pascal à Voltaire: le rôle des 'Pensées' de Pascal dans l'histoire des ideés entre 1670 et 1734, Studies on Voltaire and the Eighteenth century* 276–277 (Oxford, 1990).

15 On the image of the chain linking all phenomena, see the classic study by Arthur O Lovejoy, *The Great Chain of Being* (Cambridge, Mass., 1936, especially chapter VI).

16 Pierre Fermat (1601–65) is best remembered today for his pioneering work (with Blaise Pascal) on the calculus of probabilities.

17 Sennar, or Shinar, was the site of the tower of Babel (*Genesis*, XI, 1–9).

18 An allusion to the competing systems devised by naturalists such as Buffon and Linnæus, to which reference is made in section XLVIII below.

19 The article 'Expérimental' by d'Alembert, published in 1755 in volume V of the *Encyclopédie,* discusses the subject in very similar terms. This similarity of outlook is striking, in view of the obvious differences between the two men with regard to the value of mathematics.

20 D'Alembert had made similar comments on the philosophy of the Dark Ages in the *Preliminary Discourse* of the *Encyclopédie* (1751).

21 The reference is to Pierre Coste (1668–1747), whose edition of Montaigne's *Essais* (1724) earned him the reputation of thinking that he had written the work himself.

22 In the *Preliminary Discourse* of the *Encyclopédie,* the editors had stigmatised their contemporaries' love of false learning as a cover for ignorance, and blamed it as one of the main causes of the barbarism into which France was, they alleged, being plunged.

23 It is important to understand that Diderot had always accepted this principle, though rarely with as much emphasis as in this note and elsewhere in the *Interpretation.* Even in the *Letter on Blindness,* where the idea of monsters looms large, such aberrations are still part of the workings of the natural processes, and not something beyond rational explanation. This is, indeed, the view taken by other writers such as Buffon and Maupertuis.

24 This passage is indeed borrowed from vol. IV of Buffon's *Natural History*. It contains one of the clearest statements to be found in an eighteenth-century text of ideas which might seem to anticipate the evolutionary theories of Charles Darwin. Although he writes in volume I of the *Natural History* that species are fixed and immutable, Buffon also claims that changes in climate and diet can be responsible for significant variations between individuals of the same species. Crucially, however, he does not go on to say, any more than Diderot himself, that these changes are then perpetuated in succeeding generations. Cf. below, note 144.

25 'Baumann' is the name under which Maupertuis published the *Dissertatio*. Unlike Buffon, Maupertuis believed that changes in successive generations were the result of cumulative differences in the molecular make-up of individuals; these he ascribed not to climate or to diet, but to chance errors in the composition of the seminal fluid of the parents. Cf. below, section L, note 100.

26 Diderot's lively interest in the sexual behaviour of women is apparent not only in his own personal history, but also in novels such as *Les Bijoux indiscrets, La Religieuse* and *Jacques le fataliste,* and short stories such as *Mystification* (1768) and *Madame de La Carlière* (c.1772).

27 This theory had originated with Aristotle's *Physics,* and was still widely held at the time. It had been explained already by Maupertuis in his *Vénus physique* (1745) as well as in his *Dissertatio,* and was taken up by Buffon, as Diderot's note indicates.

28 See *Vénus physique,* chapter XVI.

29 The Classical doctrine of imitation still held sway in many intellectual circles. Contemporary French dictionaries still gave 'genius' as a synonym for 'mind', devoid of any exceptional connotations. Diderot's eulogy of the creative genius was therefore at odds with the hostility often displayed towards creativity.

30 In July 1740, Charles Bonnet's memoir on the parthenogenesis of aphids had been read to the Académie des Sciences in Paris. In 1744, Abraham Trembley had published his celebrated memoir on the fresh-water polyp. Diderot's purpose in referring to them, however obliquely, is to undermine further the belief in the need for God always to intervene in the process of creation or generation.

31 Cf. *Les Bijoux indiscrets*, I, chapter XXIX, which deals with the value of scientific hypotheses.

32 Cf. Sections V and VI above.

33 Diderot had already shown his contempt for the ignorant, unlettered section of the population in section LIII of the *Philosophical Thoughts*. In his later years, as his political writings would show, he was to become more sympathetic towards the great mass of citizens, regarding them as downtrodden and repressed by tyranny. See Anthony R Strugnell, *Diderot's Politics* (The Hague, M. Nijhoff, 1973).

34 'Rational' is used here in the sense of 'concerned with what exists in the mind', as opposed to what exists in nature.

35 The Book of Daniel (II, v. 32) refers to a monster with feet of clay seen by Nebuchadnezzar in a dream; it is destroyed by a stone 'cut out by no human hand.' Diderot's slight modification of the Biblical text again reduces the role of God.

36 Diderot develops the ideas sketched out by Bacon in Book V, chapter IV of *The Advancement and Proficiencie of Learning*, and Book I of the *Novum Organum;* in Bacon's view, the *Idolæ*, or received ideas, prevent men from acquiring a true knowledge of nature.

37 Archimedes was reputed to have said that, with his knowledge of mechanics, if he were given a place to stand he could move the world.

38 Isaac Newton had first set out the results of his experiments with the prism in *A new theory about light and colours* in 1662, though his major work in the field, the *Opticks*, did not appear until 1704. It was translated into French in 1720 by Pierre Coste (cf. note 21).

39 This division of the natural sciences is based very loosely on the scheme set out by Bacon as early as 1603 in Book III of *The Advancement and Proficiencie of Learning*.

40 An ironic reference to the 'preparation of the soul' required of devout believers, who are obliged to accept assertions founded not on experiment, but on dogma.

41 'Have Laïs, so long as Laïs does not have you.' This was, by tradition, the response given by Aristippus of Cyrenæ to those who criticised him for frequenting the prostitute Laïs. The story is also recounted in Diderot's article 'Cyrénaïque' in volume IV of the *Encyclopédie* published in 1754.

42 The story related in this and the following sections was one of Æsop's *Fables;* it was retold by La Fontaine in his own *Fables*, which were first published in 1669. These earlier versions draw the conclusion that work is its own reward, and make no mention of any lead-mine; this detail was added by Diderot to underscore the importance of developing the mineral wealth of France, and the need for co-operation between like-minded seekers after truth.

43 An echo of the practice adopted by Diderot as editor of the *Encyclopédie:* it was his custom to visit the workshops of the various craftsmen whose activities are described in the work, and to record their trades as carefully as any other subject. Many of the plates accompanying the text depict such humble toil in detail.

44 The germ of this idea is in *The Advancement and Proficiencie of Learning*, Book III, chapter V, though Bacon condemns the 'monstrous apparitions' produced by irrational beliefs. The passage is significant as an anticipation of the scientific importance of dreams to be discussed in *D'Alembert's Dream*.

45 Diderot is referring to what eighteenth-century doctors called a 'false conception', or what would now be called an ectopic pregnancy. The sac (known today as a blastotocyst, or blastoderm) contains some parts of a normal fœtus, but has no placenta, and is attached directly to the womb. As is confirmed by contemporary sources such as John Memis' *The Midwife's Pocket-Companion* (1765), medical science at that time was unable to say unequivocally whether the mola was formed spontaneously or by sexual activity. The length and technical complexity of this and other articles in this series of 'conjectures' make them similar to entries in the *Encyclopédie*. Indeed, the article 'Mole', by the surgeon Louis, which appeared in 1765 (X, 626–27), quotes at length from this section of the *Interpretation*, and refers the reader to it.

46 Here again, as in the *Letter on Blindness*, Diderot stresses the point that 'monsters' are engendered by nature, and obey the laws of nature. His purpose is not only to communicate his view that nature encompasses all of creation, but also to undermine the credulous belief in 'unnatural' diabolical creatures, such as those with which the Church frightened the faithful into compliance with its doctrines. The (unsigned) article 'Licorne' in the *Encyclopédie* gives similarly short shrift to the belief in fabulous creatures such as the unicorn.

47 The same idea was expressed by Buffon in the chapter of the *Natural History* devoted to the formation of the fœtus.

48 Buffon, in his *Theory of the Earth*, which is to be found in Book I of the *Natural History*.

49 Much of this section derives directly from the article on the aurora borealis in volume I of the *Encyclopédie* (1751), most of which is by Samuel Formey. In an editorial note at the end of the article, d'Alembert had made the same point as Diderot makes here, and had dismissed rather sharply those who ascribed the phenomenon to other causes.

50 This point, too, is made by d'Alembert in the same article.

51 D'Alembert suggests that electrical matter moves towards the north in keeping with the movement of the Earth, and escapes through the poles of a magnet. The link between the aurora and electricity in the northern parts of the globe is, he argues, a matter on which further observations would be of value(*ibid*).

52 Diderot's information on the effects of electricity is derived from various sources; of particular importance are the works of the abbé Nollet, whose *Essai sur l'électricité des corps* had appeared in 1750, and the *Experiments and Observations on Electricity* (1751) by Benjamin Franklin, which had been translated into French in 1752.

53 Diderot is very much of his time in attributing this increase in the weight of lead to the action of fire. This theory relied on the notion that a substance called 'phlogiston', which allegedly existed in all combustible bodies, was released on burning, and combined with the material being burned. It was not challenged until the 1770s, when Lavoisier's experiments began to undermine its credibility.

54 Diderot would have found much of his information on crystals in the (unsigned) articles on the subject written by the Baron d'Holbach for volume IV of the *Encyclopédie* (1754).

55 This is one of the most prescient sections of the whole work, in that it groups together phenomena now known to be caused by changes in the atomic structure

of the objects concerned. Electricity was still generally thought of merely as being induced in bodies by friction, giving them the power to attract other bodies, as when amber is rubbed: this was the definition still offered by the *Dictionnaire de Trévoux* in 1762.

56 The existence of atmospheric electricity had been discussed by Benjamin Franklin in his *Experiments and observations* (1749), and by Pierre-Charles Le Monnier in the first of his *Observations de la lune, du soleil et des étoiles fixes* in 1751. Its existence was not to be conclusively proved until the late nineteenth century, in the work of Linss.

57 That is to say, ice is not a conductor of electricity, but needs friction to induce a current to pass through it, unlike water.

58 Although the meaning of this sentence is not wholly clear, Diderot seems to be suggesting that the ice at the poles is given an electrical charge by the rotation of the earth, which has a vitreous core, and thus creates the magnetism which affects the needle of a compass.

59 Diderot's interest in mechanics had been stimulated by the publication in 1746 of Euler's *De la force de percussion*; he refers to it in the second of the *Mémoires sur différens sujets de mathématiques*(1748), which is an early version of this section of the *Interpretation*. Euler's work was also mentioned in the article 'Cordes (*Méchan.*)' which d'Alembert published in volume IV of the *Encyclopédie* in 1754. It deals with the tension of strings and the laws governing their vibration, as established by Taylor and Bernouilli, with further recent contributions from Euler and from d'Alembert himself. Diderot was therefore raising problems with which the most outstanding mathematicians of the day were also grappling.

60 The same point is made by d'Alembert in 'Cordes'.

61 The attempt to frame the laws of dynamics to cover all types of bodies flies in the face of Cartesian physics, which argued that the matter of which some bodies are composed is less dense than that of other bodies, thus creating different reactions in any given experimental situation. The point is argued in Part II of Descartes' *Principles of Philosophy* (1644).

62 In the *Encyclopédie* article 'Elasticité' (1755) d'Alembert praises this section of the *Conjectures* for being both original and valuable.

63 As so often in this work, Diderot's concern is not so much with arid technicalities as with the wider issue of the general laws of nature, on the consistency of which he insists both at the microcosmic and macrocosmic levels. The underlying guarantee of consistency is no longer, as it was in Descartes' system, the unvarying nature of God, but the nature of matter. One need hardly add that, if order is the consequence of the nature of matter, God is superfluous in the universe as conceived by Diderot.

64 It is to this section that the final *Observation* refers.

65 Diderot's belief in the theory of molecular cohesion to explain the aggregation of matter is as distinctively modern as anything in the *Interpretation*.

66 It is essential to remember that Diderot is not referring to different laws of attraction operating in the universe (since the co-existence of such laws would be incompatible with the unity of nature), but to the laws governing the attraction of certain types of molecules, which obviously vary from one substance to another, producing qualities such as hardness, softness, elasticity, etc. The point is made again at the very end of this section.

67 Cf. Note 53 above.

68 Cf. note 62 above.

69 The caves at Arcy-sur-Cure, not far from Auxerre, had yielded a large number of prehistoric animal remains. Diderot's article 'Arcy' in volume I of the *Encyclopédie* had made many of the points to which he returns, in a different context, in the present work. The subtext to these remarks is the view that the evidence garnered by science invalidated the Biblical account of creation, and was incompatible with the belief that the world was only six thousand years old. See below, note 141.

70 This section of the *Interpretation* is heavily indebted to the *Encyclopédie*, which articulates a deep practical concern for the development of trade and industry. Diderot's article 'Acier' in volume I describes in detail the process of steel-making; extensive additional information is provided in 'Fer' and 'Forge', which were written by other contributors, though they had not appeared by the time the *Interpretation* was published. 'Forge' is one of the most copiously-illustrated in the whole work, with no fewer than 52 plates.

71 Diderot is alluding here to the abbé Nollet and to Rouelle, who, since the 1730s, had been giving popular lectures on physics and chemistry during which they performed experiments. Diderot's own transcription of some of the lectures given by Rouelle which he attended in 1756–58 is reproduced in volume IX of the *Œuvres complètes de Diderot* (Paris, Hermann, 1981), p.215–39.

72 This statement sums up much of the thinking behind the *Encyclopédie*, in which the skills used in many professions and trades are revealed to the public, often for the first time.

73 Newton and Leibniz were credited with having discovered independently of each other the principles of differential calculus. Leibniz was the first to publish his discovery in 1684; Newton's first public disclosure dates from 1687, though he was known to have employed the principle in his work as early as 1669. Diderot seems always to have preferred Newton's claim to that of Leibniz.

74 The *Encyclopédie* article 'Chimie' (by the chemist Venel) states that Georg Ernst Stahl (1660–1734) published the principal ideas his master Johann Joachim Becher (1625–1682) in a style which would repel all but the most learned of readers. The remarks in this section are obviously intended as a reproach to Venel, who approves of Stahl's obscurity, on the grounds that his work might otherwise become as misunderstood in the popular mind as that of Newton.

75 Diderot's assessment of the reception his work was likely to receive was accurate: on its publication, and for years afterwards indeed, the *Interpretation* was dismissed as obscure and incomprehensible.

76 The French text at this point refers to 'l'ouvrage de Frankelin [sic]', but as several works by him had appeared by 1754, it is unclear to which of them Diderot is referring.

77 Cf. *On the Advancement and Proficiencie of Learning*, Book V, chapter II.

78 Cf.above, paragraph 6 of section XXXVI.

79 The need for some system which will allow the investigator to make sense of his results in a wider context is apparent throughout the *Interpretation*, particularly in the emphasis on discovering the general laws of nature.

80 Diderot had dwelt on the dangers of stubbornness in intellectual matters in some of his earlier works, notably the *Philosophical Thoughts* and *La Promenade du Sceptique*.

81 'Analogy' here does not mean a simple comparison, but genuine points of similarity between apparently different classes of objects. The word is used in this sense in the section of the article 'Analogie' dealing with logic and grammar (*Encyclopédie*, volume I), which is signed jointly by Du Marsais and the abbé Yvon.

82 This point is made by Bacon in *The Advancement and Proficiencie of Learning*, Book V, chapter II, with which this section has a number of incidental similarities.

83 This statement makes no sense in the revised edition of 1754: it refers to a portion of section XXXIII which appeared in the 1753 version, and which Diderot then removed. In it, he suggests that an electrical globe should be covered in gold or silver except at the poles, and friction applied while magnetised or unmagnetised needles are suspended above it, to establish whether they align themselves in a particular direction.

84 A syllogism in which one of the premises is assumed but not stated. Despite Diderot's assertion, it is unclear that two different experiments based on the same premise will necessarily lead to results which are identical in all respects.

85 Diderot's desire to discover the underlying unity of nature, and his conviction that such unity must exist, is manifest in this section.

86 As in earlier sections of the *Interpretation,* mathematics is not seen as a way of achieving this perfect state of knowledge; Diderot merely uses it as an analogy to indicate that experimental science should aim as far as possible to achieve the degree of completeness found in mathematical demonstrations.

87 Cf. the end of section X of the *Interpretation.*

88 The importance of chance in experimental science had been stressed by Bacon in the *Advancement*, Book V, chapter II.

89 Bacon had devoted much of Book V, chapter IV of the *Advancement* to a consideration of the *Idolaes [sic]*, or impediments to understanding; among these he includes the *'Idola Theatri, or depraved Theories or Philosophies, and perverse Laws of Demonstrations.'*

90 Diderot uses the word 'méthodistes' to designate Linnæus and his followers. In the *Interpretation*, the term must be taken to mean 'those who classify or arrange according to a particular method or scheme' (OED). This sense was current in English in the mid-eighteenth century, before it became inextricably associated with non-conformist Protestantism.

91 Essentially the same points had already been made by Buffon in the volumes of the *Natural History* devoted to the history of man (1749). The chapters devoted to monkeys in the volumes dealing with the history of quadrupeds (1753–67) contain further reflections on the subject.

92 Carl von Linné (1707–78), known from the latinised form of his name as Linnæus, is regarded as the founder of modern systematic botany; he was the first to lay down the principles for defining genera and species, and to use specific names in a uniform and consistent way. His *Fauna Suecica* appeared in 1746.

93 Diderot reproduces accurately the substance of the preface to the *Fauna*, though he omits Linnæus's eulogy of human reason, which significantly modifies the statements quoted here.

94 See above, Diderot's first note to section XII. The *Dissertatio* to which Diderot alludes here (and from which he sometimes quotes directly) was published in a very small number of copies, and is virtually unobtainable. The French version, the *Essai sur la formation des corps organisés* (1754), which Diderot also uses, was reprinted in a number of editions of Maupertuis's works.

95 This is a reference to the physics of Descartes, as set out in his *Principia philosophiæ [Principles of Philosophy]*(1644).

96 While this may be a reference to the physics of Newton, it could also refer to d'Alembert who, in the article 'Matière' of the *Encyclopédie* (1751), had stated that impenetrability, motility and divisibility are the essential qualities of matter.

97 A direct reference to Newton's *Principia mathematica* (1687).

98 The term 'plastic nature', which was widely used, was taken from the works of two English neo-platonist writers, Henry More's *Enchiridion metaphysicum* (1671) and Ralph Cudworth's *True intellectual system of the universe* (1678). The essence of this doctrine is that it supposes that God is separate from matter, which has its own laws of organisation ordained by Him.

99 A reference to the theory of *emboîtement* or pre-formation which was widely held in the eighteenth century, in opposition to that of epigenesis, or the continuous creation of new organisms.

100 In the edition of his works published in Lyon in 1756, Maupertuis responded somewhat irritably to Diderot's comments on his work, and repudiated the use of the term 'sensitive soul' as irrelevant to his own ideas. Diderot seems to be mixing up the traditional division of the soul into three parts, the sensitive, the rational and the vegetative: the sensitive soul was found only in animals, and was not common to all minds of matter, as he alleges.

101 Any accusation of materialism and atheism would have greatly troubled the devout Maupertuis. In the foreword to his *Essai de Cosmologie* (1750), he maintains that the sheer abundance of the wonders of nature compel belief in God.

102 Diderot is careful to report accurately the substance (and sometimes the actual words) of Maupertuis' *Essai sur la formation des corps organisés*.

103 This is the first occurrence in Diderot's writings of an image to which he will return in *D'Alembert's Dream*. The essence of the idea is to be found in Maupertuis' *Essai*.

104 See above, sections XI and XII. Maupertuis had explained the existence of new species by a succession of chance mutations, and refers to each divergence from the original as an 'error'. Diderot makes no such judgement on the workings of nature. Modern geneticists are inclined, on the basis of DNA evidence, to accept that all life originated in one original act of creation.

105 This is a variation on the problem raised in section XXXVI in relation to the behaviour of molecules in mixed substances.

106 'It appears that when all the perceptions of the elements are brought together, the result is one single perception, which is much stronger and more perfect than any of the elementary perceptions. This perception may stand in the same relationship to these perceptions as the organised body to the element. Once each element has united with the others, has blended its perceptions with theirs, and has lost the feeling of being a particular self, the memory of the primitive state of the elements ceases, and our origins become entirely lost to us.' Diderot's reason for quoting the original Latin of Maupertuis's text is no doubt to avoid accusations of misrepresentation in this crucial matter.

107 This remark fooled nobody, least of all Maupertuis, who, in his reply to Diderot, accused him of wanting to draw exactly these consequences from the *Essai* (see above, note 97).

108 Cf. sections VI and XI above, and section LVIII, question 1.

109 This is the crux of Diderot's argument. If the world is effectively the totality of all created things, and God is by definition infinite, then God and the world cannot be separate. This pantheistic outlook goes back to Antiquity, and can be found in the works of the Stoics among others; it had been given a new impetus by the *Ethics* of Spinoza (1677). The Church condemned it as a purely atheistic doctrine. Maupertuis' response was to reject any link with Spinozism in this sense, and to maintain that the universe is not a whole; he does not explore the implications of this denial.

110 Diderot's purpose in analysing Maupertuis' work so closely is apparent here. Maupertuis (1698–1759) had been President of the Royal Academy of Sciences in Berlin, and was celebrated for his pioneering work in astronomy, geometry and the theory of reproduction. He was therefore an important figure in the debate on the nature of the universe, and (no doubt unintentionally) lent considerable weight to the arguments deployed by the relatively unknown Diderot.

111 'Organic molecules' (an expression derived from Buffon) are the elementary parts of living beings, the smallest units capable of sentient life.

112 This is the clearest statement in the *Interpretation* of Diderot's belief that matter is self-organising. It underlies much of what we find elsewhere in the work, and looks forward to the more complex theories put forward in *D'Alembert's Dream* and *Rameau's Nephew*.

113 See above, sections XVIII and XXII.

114 This is a return to the attack on mathematics launched in the early sections of the *Interpretation*. For once, Diderot does not echo Bacon, who extolled the advantages of mathematics in measuring quantities (see *On the Advancement and Proficiencie of Learning*, Book III, chapter VI).

115 Cf. above, sections XXIII, XXXI and XLIII.

116 But see above, section XIII.

117 Diderot may be thinking of Réaumur's *Mémoires pour servir à l'histoire des insectes*, which had been published in six volumes between 1734 and 1742. But he also has in mind, more generally, the 'methodists' whom he had taken to task in sections XLVIII and XLIX.

118 These names are not chosen at random, but are the expression of Diderot's frustration at the hostility attending the publication of any work which challenged ecclesiastical orthodoxy. Montesquieu's *De l'Esprit des Loix [On the Spirit of the Laws]*(1748) had been viciously attacked in the Jansenist periodical *Les nouvelles ecclésiastiques* for its alleged Spinozist tendencies. Its hostile reception had become something of a *cause célèbre* with the encyclopedists, who praised it frequently in various articles. On Montesquieu's death in 1755, a sixteen-page eulogy of him was added to volume V of the *Encyclopédie*. Diderot was the only one of the *Philosophes* to attend his funeral. The *Lettres à un Américain* (1751) by another Jansenist, Lelarge de Lignac, had attacked Buffon's *Natural History*.

119 As this outburst indicates, Diderot himself had not been spared similar criticisms. The *Journal de Trévoux*, a Jesuit publication edited by Guillaume-François Berthier, had repeatedly savaged the *Encyclopédie*, which it saw as a dangerously free-thinking rival to its own *Dictionnaire de Trévoux*. The Jesuits had also been instrumental in securing the suspension of the *Encyclopédie* for several months in 1752; in 1759, following its condemnation by the Pope, it was to be officially suppressed for a number of years. On contemporary reactions to the *Encyclopédie*, see John Lough, *Essays on the* Encyclopédie *of Diderot and D'Alembert* (Oxford, 1968), chapters V and VI.

120 Charles Pinot-Duclos (1704–72) was the Permanent Secretary of the Académie Française, and the author of several novels and works of history.

Jean Le Rond d'Alembert (1717–83) was, in addition to being the co-editor with Diderot of the *Encyclopédie*, perhaps the most outstanding mathematician in Europe at that time.

Jean-Jacques Rousseau (1712–78) had contributed articles on musicology to the *Encyclopédie*, and was already noted for his attacks on the harm done to mankind by the arts and sciences. He was soon to break publicly with Diderot and the

Philosophes, with whose belief in material progress he increasingly disagreed.

François-Marie Arouet de Voltaire (1694–1778) was the leading figure in the *Philosophe* camp; he was mainly known at this time for his tragedies and works of history, including the *Siècle de Louis XIV* (1751), mentioned later in the paragraph. On Maupertuis, see above, note 110.

In addition to *De l'Esprit des Loix*, Charles-Louis de Secondat, baron de Montesquieu (1689–1755) had written the *Lettres persanes [Persian Letters]* (1721), together with a number of works on the relationship between customs and climate, and on the general influences shaping human conduct.

Georges-Louis Leclerc, comte de Buffon (1707–88) began publishing his monumental *Histoire naturelle* in 1749; the final volumes, completed by other hands, did not appear until 1804.

Louis-Jean-Marie Daubenton (1716–1800) was one of the greatest naturalists and anatomists of his time.

121 In other words, that the universe has no beginning and no end, in contradiction to what is said in the Bible.

122 Saunderson, in the *Letter on the Blind*, makes this point in a different way: because he cannot see them, he refuses to concede that the wonders of nature prove the existence of God. The doctrine that we cannot know that something exists if we have no sensory knowledge of it was the core of Bishop Berkeley's philosophy, as set out in his *Dialogues of Hylas and Philonous* (1713). Diderot had already discussed Berkeley's ideas in *La Promenade du Sceptique* (1747), but had been unable to refute them to his own satisfaction.

123 This description of the approach adopted by the interpreter of nature differs from that of the systematising rationalist: the interpreter uses experimental evidence, rather than reason alone, to take him as far in his investigations as possible; on that basis, he then formulates more general conclusions, stopping only where human knowledge cannot take him, namely into the realms of theology.

124 This rejection of final causes was a hallmark of *Philosophe* thought. In his article 'Causes finales' in volume II of the *Encyclopédie* (1752), d'Alembert had agreed with Bacon that the search for final causes was futile.

125 That is, theology 'based upon reasoning from natural facts apart from revelation' (OED).

126 Hence, any attempt to determine *why* nature operates as it does must inevitably be speculative, and lead use away from the study of *how* it operates, which is the true realm of the natural scientist.

127 Exupère-Joseph Bertin (1712–81) carried out a series of experiments on the lymphatic system and on the blood vessels. In medical science, 'anastomosis' designates communication between two distinct bodily entities of any kind, such as arteries and veins, or, as here, between two different arterial systems. The article on the subject which the Baron de Haller contributed to volume I of the *Encyclopédie* provided Diderot with a wealth of information.

128 'The heavens are telling the glory of God' (Psalms, XIX, v.1).

129 Galen (2nd century AD) long remained the standard source of medical knowledge. His treatise on the human body contained a number of statements which seemed to indicate a finalist interpretation of God's purpose in creating man in His own image. Hermann Boerhaave (1668–1738) was a professor of medicine at Leyden; he was the inspiration behind the 'man is a machine' thesis of his pupil La Mettrie, who translated some of his 'works into French, though none exists with this title. The Swiss physiologist Albrecht von Haller (1708–77) was also a

pupil of Boerhaave; he secured a European reputation for his research in medicine, botany and anatomy. His major work, the *Elementa physiologiæ corporis humani*, did not begin to appear until 1757, but Diderot may be basing his remarks on his short treatise *Primæ linæ physiologiæ* (1751). Despite Haller's genius for rigorous scientific experimentation, his letters show him to have been a somewhat naïve finalist in religious matters.

130 For Bacon's views on the preconceptions which threaten to deform the judgement, see above, note 89. The attack on 'common sense' had been set out in Descartes' *Discours de la méthode [Discourse on Method]* (1637), in which he comments that it is not enough to have good sense; one must also know how to use it.

131 *Ecclesiastes*, I, 9.

132 A reference to the *Lettres de Leibnitz et de Clarke* (1717), in which Leibniz mentions that a friend of his had tried unsuccessfully to find two exactly similar leaves in a garden. Leibniz had not laid stress on the observer's *perceiving* the leaves to be of the same shade, and Diderot's emphasis derives from his insistence on the importance of sensory observations for the empirical scientist.

133 Cf. above, section L, note 104. It should be noted that Diderot refers only to the process of change taking place over time, and does not venture to explain it; that is to say, he does not consider the causal development of species which lies at the heart of the theory of evolution.

134 In his article 'Goût' ['Taste'] in volume VII of the *Encyclopédie* (1757), Voltaire was to assert that good taste consisted of admiring that which pleased cultivated minds. Diderot's views on taste were to become rather more permissive in later years, particularly when he became interested in art criticism at the end of the 1750s.

135 This was the first of Diderot's works to conclude with questions, and indicates the open-ended, enquiring nature of the *Interpretation*.

136 Diderot refuses consistently to entertain the idea that all creation could have originated from only one type of matter; the development of his thought in this section recalls what was said in section XXXVI, paragraph 5 and 6, on the heterogeneity of matter resulting from the mixing of different kinds of molecules.

137 Science was to echo Diderot's views in this respect for another century or more. In 1884, the OED was still defining an atom as 'a theoretical body, so infinitely small as to be incapable of further division.'

138 Cf. section LVII.

139 Diderot's doubts are the symptom of a perplexity from which he was never wholly to free his mind: is what we see now merely a moment of apparent stasis in the infinite chaos of nature, from which we can draw no conclusions, or does it represent what nature has always been, and will always be, as a result of the operation of its laws? On balance, he inclines to the latter view, but can never wholly rid himself of the former.

140 Prescient thoughts such as this put one in mind of recent debates on the fate of the dinosaurs which, many scientists have argued, mutated into birds as a result of catastrophic environmental changes.

141 As an indication of the boldness of these speculations on the development of species over millions of years, it should be remembered that the Church still accepted the calculation, based on the Bible, that the age of the world was 6,000 years (see Diderot's article 'Age (Mythologie)' in volume I of the *Encyclopédie*). Cf. above, note 69.

142 The division of matter into these two categories, a thesis which Diderot rejects here, is put forward (though not in such stark terms) by Buffon in volume I of his *Natural History* (1749).

143 Although the issues raised here are put forward tentatively in the form of questions, it is clear that Diderot envisages matter as containing its own energy, as self-propelled, and that he rejects the idea that any matter can be truly dead. There is thus no need for God to give any impetus to matter, and the unity of nature in purely material terms is strongly implied. These ideas will be further explored in *D'Alembert's Dream*.

144 The matrix theory had been put forward by Buffon in his volume on the history of animals to explain why each species has a particular form. Despite his insistence that species do not change fundamentally over time, the matrix theory sits uneasily with a belief in the variations produced in individuals as a result of environmental changes (cf. above, note 24): each new variation would require its own matrix, or would imply that each matrix is infinitely variable in unpredictable ways. Even if this somewhat elastic definition of a matrix is accepted, it does not account for the fact, acknowledged by Buffon (in *The Epochs of Nature* in volume I) that the remains of ancient species show that they were generally larger than their modern descendants; such generalised changes, whether due to climate or to diet, would imply a permanent change in the matrix, for which his theories cannot readily account.

145 This problem is closely related to the matters discussed in paragraphs 5 and 6 of section XXXVI.

146 Questions 14 and 15 cannot be answered in the static, fixed terms of Buffon's theories. As Diderot indicates in the closing paragraph, they represent the ultimate theoretical problems which scientists should leave aside until the practical difficulties facing mankind have been dealt with.

147 In arguing that Cartesian philosophy (unlike that of Newton perhaps) is unable to explain the variability of action in a vacuum, Diderot is underlining the inadequacy of traditional physics to cope with the discoveries of modern science. A similar attack on the weaknesses of Cartesian science had already been made by Voltaire in his *Letters concerning the English Nation* (1733), and by d'Alembert in the article 'Cartésianisme' in volume II of the *Encyclopédie* (1752). The article was translated into English as 'Des Cartes' Philosophy', and published in the *Select Essays from the Encyclopedy* (London, 1772), p.316–72.

D'Alembert's Dream

D'Alembert's Dream[1]

The Speakers: *d'Alembert, Madamoiselle de l'Espinasse and Doctor Bordeu.*

Bordeu: Well! What's been happening now? Is he ill?

Mlle. de l'Espinasse: I'm afraid so; he had the most restless night.

Bordeu: Is he awake?

Mlle. de l'Espinasse: Not yet.

Bordeu (after going up to d'Alembert's bed and feeling his pulse and his skin): It'll be nothing.

Mlle. de l'Espinasse: You think so?

Bordeu: I'm sure of it. His pulse is good … somewhat weak … his skin moist … his breathing easy.

Mlle. de l'Espinasse: Is there anything to be done for him?

Bordeu: Nothing.

Mlle. de l'Espinasse: So much the better, for he hates medicines.

Bordeu: And so do I. What did he eat for supper?

Mlle. de l'Espinasse: He wouldn't take anything. I don't know where he had been spending the evening, but he seemed worried when he came back.

Bordeu: Just a slight touch of fever that won't have any ill effects.

Mlle. de l'Espinasse: When he got home, he put on his dressing-gown and nightcap and flung himself into his armchair, where he dozed.

Bordeu: Sleep is good anywhere, but he would have been better in bed.

Mlle. de l'Espinasse: He was angry with Antoine for telling him so; he had to be worried for half an hour to get him to bed.

Bordeu: That happens to me every day, although I'm in good health.

Mlle. de l'Espinasse: When he was in bed, instead of resting as usual, for he sleeps like a child, he began to toss and turn, to stretch out his arms, throw off his covers and talk aloud.

Bordeu: And what was he talking about? Geometry?

Mlle. de l'Espinasse: No; it really sounded like delirium. To begin with, a lot of nonsense about vibrating strings and sensitive fibres. It seemed so crazy to me that I resolved not to leave him alone all night, and not knowing what else to do I drew up a little table to the foot of his bed, and began to write down all I could make out of his ramblings.

Bordeu: A good notion, and typical of you. Can I have a look at it?

Mlle. de l'Espinasse: Surely; but I'll stake my life you won't understand a thing.

Bordeu: Perhaps I may.

Mlle. de l'Espinasse: Are you ready, Doctor?

Bordeu: Yes.

Mlle. de l'Espinasse: Listen. 'A living point... no, I'm wrong. First nothing, then a living point. To this living point is applied another, and yet another; and the result of these successive increments is a being that has unity, for I cannot doubt my own unity.' As he said this, he felt himself all over. 'But how did this unity come to be?' Oh, my friend, I said to him, what does that matter to you? Go to sleep. He was silent for a moment, but began again as if speaking to someone: 'I tell you, philosopher, I can understand an aggregate or tissue of tiny sensitive beings, but not an animal ... a whole a system, an individual, having consciousness of its unity! I can't accept that, no, I can't accept it.' Doctor, can you make anything of it?

Bordeu: A great deal.

Mlle. de l'Espinasse: Well, you're lucky. 'Perhaps my difficulty comes from a mistaken idea.'

Bordeu: Are you speaking yourself?

Mlle. de l'Espinasse: No, that's the dreamer. I'll go on. He added, apostrophizing himself: 'Take care, friend d'Alembert, you are assuming only contiguity where there exists continuity... yes, he's clever enough to tell me that. And how is this continuity formed? That won't offer any difficulty to him. As one drop of mercury coalesces with another drop of mercury, so one living and sensitive particle coalesces with another living and sensitive particle.[2] First there were two drops, after the contact there is only one. Before assimilation there were two particles, afterwards there was only one... sensitiveness becomes a common property of the common mass. And indeed why not? I may imagine the animal fibre divided up into as many sections as I please, but that fibre will be continuous, will be a whole, yes, a whole. Continuity arises from the contact of two perfectly homogeneous particles; and this constitutes the most complete union, cohesion, combination, identity that can be imagined ... yes, philosopher, if these particles are elementary and simple; but what if they are aggregates, what if they are compound? They will combine none the less, and in consequence become united, continuous ... and then there is continual action and reaction. It is certain that contact between two living particles is quite different from contiguity between two inert masses. ... Let that pass; it might be possible to start a quarrel with you on that point; but I don't care to do so, I don't like carping. Let's go back to where we were. A thread of purest gold, I remember, was one comparison he used; a homogeneous network between the particles of which others thrust themselves and form, it may be, another unified network, a tissue of sensitive matter; contact involving assimilation; sensitiveness, active in one case, inert in another, which is communicated like motion, not to mention that, as he very well put it, there must be a difference between the contact of two sensitive particles and the contact of two that are not sensitive; and wherein can that difference lie? ... a continual action and reaction ... and this action and reaction having a particular character. Everything then, concurs to produce a sort of unity which exists only in the animal. Well! if that's not truth it's very like it.'[3] Doctor, you're laughing; can you see any sense in this?

Bordeu: A great deal.

Mlle. de l'Espinasse: Then he's not mad?

Bordeu: By no means.

Mlle. de l'Espinasse: After this preamble he began to cry: 'Mademoiselle de l'Espinasse! Mademoiselle de l'Espinasse!' 'What do you want?' 'Have you sometimes seen a swarm of bees escaping from their hive? ... The world, or the general mass of matter, is the hive. Have you seen them go and form, at the end of the branch of a tree, a long cluster of little winged animals, all clinging to one another by their feet? This cluster is a being, an individual, an animal of sorts. But such clusters should all be alike. Yes, if he accepted only a single homogeneous matter. Have you seen them?' 'Yes, I've seen them.'

'You've seen them? 'Yes, my friend, I tell you, yes.' 'If one of these bees should take a fancy to nip, in some way, the next bee it's attached to, what do you think will happen? Tell me.' 'I don't know.' 'Go on, tell me. You don't know then, but the philosopher knows well enough. If you ever see him – and you may or may not see him, for he promised me – he will tell you that this bee will nip the next; that, throughout the cluster, there will be aroused as many sensations as there are little animals; that the whole will be disturbed, will stir, will change its position and its shape; that a noise will arise, little cries, and that anyone who had never seen a similar cluster in formation would be inclined to take it for an animal with five or six hundred heads and a thousand or twelve hundred wings.' Well, Doctor?

Bordeu: Well, do you know, that's a very fine dream, and you were quite right to take it down.

Mlle. de l'Espinasse: Are you dreaming too?

Bordeu: So far from it, that I'd almost undertake to tell you how it goes on.

Mlle. de l'Espinasse: I defy you to.

Bordeu: You defy me?

Mlle. de l'Espinasse: Yes.

Bordeu: And if I get it right?

Mlle. de l'Espinasse: If you get it right I promise... I promise... to take you for the greatest madman on earth.

Bordeu: Look at your paper and listen to me. 'A man who took this cluster to be an animal would be wrong.' But, Mademoiselle, I presume he went on addressing you. 'Would you like him to judge more sanely? Would you like to transform the cluster of bees into one single animal? Modify a little the feet by which they cling together; make them continuous instead of contiguous. Between this new condition of the cluster and the former, there is certainly a marked difference; and what can that difference be, if not that now it is a whole, a single animal, whereas before it was a collection of animals? All our organs...'

Mlle. de l'Espinasse: All our organs!

Bordeu: To one who has practised medicine and made a few observations ...

Mlle. de l'Espinasse: Next?

Bordeu: Next? ... 'Are just separate animals held together by the law of continuity in a general sympathy, unity and identity.'

Mlle. de l'Espinasse: I'm dumbfounded! You've got it almost word for word. Now I can proclaim to all the world that there's no difference between a waking doctor and a dreaming philosopher.

Bordeu: That was already suspected. Is that the whole of it?

Mlle. de l'Espinasse: Oh no, not nearly. After your, or his, ravings, he said: 'Mademoiselle?' 'Yes, my friend?' 'Come here... nearer, nearer. I want you to do something.' 'What is it?' 'Take this cluster, here it is, you're sure it's there? now, let's make an experiment.' 'What experiment?' 'Take your scissors: do they cut well?' 'Perfectly.' 'Go up gently, very gently, and separate these bees, but be careful not to divide them through the middle of the body; cut just where they're joined on to one another by the feet. Don't be afraid. You may hurt them a little, but you won't kill them. Very good, you're as skilful as a fairy. Do you see how they fly apart on every side? They fly one by one, in twos, in threes. What a lot of them there are! If you've understood me ... you're sure you've understood me?' 'Quite sure.' 'Now suppose... suppose...' On my word, Doctor, I understood so little of what I was writing, he was speaking so softly, this part of my paper is so much scribbled over, that I can't read it.

Bordeu: I'll fill in the gaps, if you like.

Mlle. de l'Espinasse If you can.

Bordeu: Nothing easier. 'Suppose these bees to be so tiny, that their organisms always escaped the coarse blade of your scissors: you could go on dividing as much as you pleased, without killing one of them, and this whole, composed of imperceptible bees, would really be a polypus that you could destroy only by crushing. The difference between the cluster of continuous bees and the cluster of contiguous bees is precisely that existing between ordinary animals like ourselves or the fishes on the one hand and worms, serpents and polypous animals; moreover the whole of this theory undergoes further modifications.' (*Here Mlle. de l'Espinasse gets up suddenly and pulls the bell- cord.*) Gently, gently Mademoiselle, you will wake him, and he needs rest.

Mlle de l'Espinasse: I'm so bewildered I never thought of that. (*To the servant who enters*) Which of you went to the doctor's?

Servant: I did, Mademoiselle.

Mlle. de l'Espinasse: How long ago?

Servant: I've not been back an hour.

Mlle. de l'Espinasse: Did you take anything there?

Servant: Nothing.

Mlle. de l'Espinasse: No paper?

Servant: None.

Mlle. de l'Espinasse: All right, you may go ... I can't get over it! Look here, Doctor, I suspected one of them of letting you see my scribble.

Bordeu: I assure you that's not so.

Mlle. de l'Espinasse: Now that I've discovered your gift, you'll be a great help to me socially. His dream talk didn't end there.

Bordeu: All the better.

Mlle. de l'Espinasse: You see nothing to worry about in that?

Bordeu: Nothing at all.

Mlle. de l'Espinasse: He went on... 'Well, then, philosopher, do you imagine polypi of every sort, even human polypi? ... But nature shows us none.'

Bordeu: He did not know of the two girls who were joined together by their heads, shoulders, backs, buttocks and thighs, who lived thus joined together to the age of twenty-two, and died within a few minutes of each other. Then what did he say?

Mlle. de l'Espinasse: The sort of things you hear only in a madhouse. He said: 'It has happened or else it will happen. And who knows the state of things on other planets?'

Bordeu: Perhaps there's no need to go so far.

Mlle. de l'Espinasse: On Jupiter or on Saturn, human polypi! Males splitting up into males, females into females, it's an amusing notion.' Thereupon he burst into fits of laughter that were quite terrifying. 'Man splitting up into an infinite number of atomic men, that can be wrapped between sheets of paper like insects' eggs, that spin their cocoons, remain as chrysalides for a certain time, then break through their cocoons and escape like butterflies, a society of men formed and a whole province peopled out of the fragments of a single man, it's quite delightful to imagine.' And then he burst out laughing again. 'If, somewhere or other, man splits up into an infinite number of human animalcules, death must be less dreaded; the loss of a man is so easily repaired that it ought to cause very little grief.'

Bordeu: This extravagant hypothesis is almost the true story of all the species of animals which exist now and which are to come. If man does not split up into an infinite number of men, at any rate he splits up into an infinite number of animalcules, whose metamorphoses and whose future and final organization cannot be foreseen. Who knows if this is not the nursery of a second generation of beings, separated from this generation by an inconceivable interval of centuries and successive developments?

Mlle. de l'Espinasse: What are you muttering away there, Doctor?

Bordeu: Nothing, nothing, I was just dreaming on my own account. Go on reading, Mademoiselle.

Mlle. de l'Espinasse: 'Everything considered, however, I prefer our way of renewing the population,' he added. 'Philosopher, you who know what happens here, there and everywhere, tell me, doesn't the dissolution of different parts produce men of different characters? The brain, the heart, the chest, the feet, the hands, the testicles. ... Oh! how this simplifies morality! A man born, a woman brought forth.' Doctor, you'll allow me to pass over this.' 'A warm chamber, lined with little packets, on each packet a label: *warriors, magistrates, philosophers, poets,* packet of *courtiers,* packet of *whores,* packet of *kings.'*

Bordeu: This is very merry and very mad. This is a dream indeed, and a vision that calls up certain strange phenomena to my mind.

Mlle. de l'Espinasse: Then he began to mutter something or other about grains, strips of flesh put to macerate in water, different and successive races of creatures that he beheld being born and passing away. With his right hand he had imitated the tube of a microscope, and with his left, I think, the mouth of a vessel. He was looking into this vessel through the tube and saying: 'Voltaire can make fun of it as much as he likes, but the "Eel-man"[4] is right; I believe my eyes; I can see them; what a lot there are! how they come and go, how they wriggle!' The vessel in which he perceived so many short-lived generations, he compared to the Universe: he saw the history of the world in a drop of water. This idea seemed a tremendous one to him; it appeared to fit in perfectly with sound philosophy, which studies great bodies in little ones. He said: 'In Needham's[4] drop of water, everything occurs and passes away in the twinkling of an eye. In the world, the same phenomenon lasts a little longer; but what is our duration compared with the eternity of time? Less than the drop I have taken up on the point of a needle compared with the limitless space that surrounds me. An unbounded series of animalcules in the fermenting atom, the same unbounded series of animalcules in this other atom[5] that is called the Earth. Who knows what races of animals have preceded us? Who know what races of animals will come after ours? Everything changes and everything passes away, only the whole endures. The world is forever beginning and ending; each instant is its first and its last; it never has had, it never will have, other beginning or end.[6] In this vast ocean of matter, not one molecule is like another, no molecule is for one moment like itself. *Rerum novus nascitur ordo** is eternally inscribed upon it.' Then he added with a sigh: 'O the vanity of our thoughts! O the poverty of fame and of all our labours! O wretchedness! O the brief scope of our understanding! Nothing is solid save drinking, eating, living, loving and sleeping. Mademoiselle de l'Espinasse, where are you?' 'Here I am.' Then his face flushed. I wanted to feel his pulse, but I did not know where he had hidden his hand. He appeared to undergo a convulsive movement. His mouth was half-open, his breathing hurried: he heaved a deep sigh, then a weaker and still deeper sigh; he turned his head over on his pillow and fell asleep. I looked at him attentively, and was much moved without knowing why; my heart was throbbing, and it wasn't from fear. After a few minutes, I saw a slight smile flit across his lips; he whispered: 'On a planet where men multiplied after the fashion of fishes, where the spawn of a man in contact with a woman's spawn ... then I'd regret it less. Nothing should be lost that might be useful. Mademoiselle, if it could be collected, sealed in a flask and sent very early to Needham.' Doctor don't you call this madness?

Bordeu: When he was near you, assuredly!

Mlle. de l'Espinasse: Near me, away from me, it's all the same; you don't know what you're talking about. I had hoped that the rest of the night would be quiet.

Bordeu: Such is usually the result.

Mlle. de l'Espinasse: Not at all; about two in the morning he harked back to his drop of water, calling it a mi... cro...

Bordeu: A microcosm.

Mlle. de l'Espinasse: That was the word he used. He was admiring the wisdom of the ancient philosophers. He was saying, or making his philosopher say, I don't know which: 'If when Epicurus[7] maintained that the earth contained the germs of everything, and that the animal species was a product of fermentation, he had proposed to show an illustration on a small scale of what happened on a large scale at the beginning of all time, what would have been the answer? And you have such an illustration before your eyes, and it teaches you nothing. Who knows whether fermentation and its products are exhausted? Who knows what point we have reached in the succession of these generations of animals? Who knows whether that deformed biped, a

* A new order of things comes into being.

mere four feet high, who is still called a man in the region of the Pole and who would quickly lose the name by growing a little more deformed, does not represent a disappearing species? Who knows if this is not the case with all species of animals? Who knows whether everything is not tending to be reduced to one vast, inert, motionless sediment? Who knows how long that inertia will endure? Who knows what new race may spring up again from such a great agglomeration of sensitive and living points? Why not one single animal? What was the elephant originally? Maybe the same huge animal that we know today, maybe an atom – both are equally possible; you need assume only motion and the varied properties of matter. The elephant, that huge organized mass, a sudden product of fermentation! Why not? There is less difference between that great quadruped and its first matrix than between the tiny worm and the particle of flour whence it sprang; but the worm is only a worm… that is, its smallness, by concealing its organization from you, takes away the element of wonder. Life, sensitivity, therein lies the miracle; and that miracle is one no longer. When once I have seen inert matter attain the state of feeling, of sensitivity, there is nothing left that can astonish me. What a comparison! A small number of elements in a state of ferment in the hollow of my hand, and this immense reservoir of divers elements scattered through the bowels of the earth, over its surface, on the bosom of the sea, in the void of the air! And yet, since the same causes persist, why have their effects ceased? Why do we no longer see the bull pierce the earth with his horn, press his hoofs against the soil, and struggle to disengage his ponderous body from it? Let the present race of existing creatures pass away; leave the great inert sediment to work for a few million centuries. It may be that the renewal of species takes ten times longer than their allotted span of life. Wait, and do not give a hasty judgment on the great work of nature. You have two great phenomena; the transition from the state of inertia to the state of sensitivity, and spontaneous generation;[8] let these suffice you; draw correct conclusions from them, and in an order of things which allows no absolute degree of greatness or smallness, permanence or transience, avoid the sophistry of the ephemeral.' Doctor, what is this sophistry of the ephemeral?

Bordeu: That of a transient being who believes in the immortality of things.

Mlle. de l'Espinasse: Fontenelle's rose, saying that within the memory of a rose no gardener had been known to die?

Bordeu: Precisely; that is graceful and profound.

Mlle. de l'Espinasse: Why don't your philosophers express themselves with the grace he does? We should understand them then.

Bordeu: Frankly, I do not know if that frivolous tone suits serious subjects.

Mlle. de l'Espinasse: What do you call a serious subject?

Bordeu: Why, the general sensitivity of matter, the formation of the sentient being, its unity, the origin of animals, their duration, and all the questions connected with these.

Mlle. de l'Espinasse: Well, I call those crazy questions, about which one may dream when one is asleep, but which no man of sense will trouble about in his waking hours.

Bordeu: And why so, if you please?

Mlle. de l'Espinasse: Because some are so obvious that it's useless to seek their explanation, others so obscure that they can't possibly be understood, and all completely useless.

Bordeu: Do you think it a matter of indifference, Mademoiselle, whether one denies or accepts the existence of a Supreme Intelligence.

Mlle. de l'Espinasse: No.

Bordeu: Do you think one can come to a decision about the Supreme Intelligence without knowing what opinion to hold as to the eternity of matter, its properties, the distinction between the two substances, the nature of man and the production of animals?

Mlle. de l'Espinasse: No.

Bordeu: So, then, these questions are not as idle as you said they were.

Mlle. de l'Espinasse: But what does their importance matter to me, if I cannot solve them?

Bordeu: And how can you do that if you won't examine them? But may I ask you which are those problems which you find so plain that examination of them appears to you superfluous?

Mlle. de l'Espinasse: The question of my unity, of my individual identity, for instance. Heavens, it seems to me there's no need of so much talk to tell me that I am myself, that I have always been myself and shall never be anybody else.

Bordeu: No doubt the fact is plain, but the reason for the fact is by no means so, especially on the hypothesis of those who only admit a single substance and who explain the formation of man, or animals in general, by a series of contacts between sensitive particles. Each sensitive particle had its individual identity before the contact; but how did it lose it, and how from all these losses did there result the consciousness of a whole?

Mlle. de l'Espinasse: It seems to me that contact, in itself, is enough. Here's an experiment I've made a hundred times … but wait, I must go and see what's happening behind those curtains … he's asleep. … When I lay my hand on my thigh, I can clearly feel at first that my hand is not my thigh, but some time after, when both are equally warm, I can no longer distinguish between them; the limits of the two parts of my body become blended and make only one.

Bordeu: Yes, until one or the other receives a prick; then the distinction reappears. So, then, there is something in you that knows whether it is your hand or your thigh that has been pricked, and that something is not your foot, nor even your pricked hand – the hand suffers, but the other thing knows and does not suffer.

Mlle. de l'Espinasse: Why, I think it's my head.

Bordeu: Your whole head?

Mlle. de l'Espinasse: No; look, Doctor, I shall explain myself by means of a comparison, since comparisons make up almost the whole argument for women and poets. Imagine a spider…

d'Alembert: Who's there? Is it you, Mademoiselle de l'Espinasse?

Mlle. de l'Espinasse: Hush, hush. (*Mlle. de l'Espinasse and the doctor are silent for some time, then Mlle. de l'Espinasse says softly*): I think he's gone to sleep again.

Bordeu: No, I fancy I hear something.

Mlle. de l'Espinasse: You're right; is he beginning to dream again?

Bordeu: Let's listen.

d'Alembert: Why am I what I am? Because it was inevitable I should be. Here, yes, but elsewhere? At the Pole, below the Equator, on Saturn? If a distance of a few thousand leagues can alter my species, what will be the effect of an interval of many thousand times the world's diameter? And if all is in perpetual flux, as the spectacle of the Universe everywhere shows me, what may not be produced here and elsewhere by the lapse and vicissitudes of several million centuries? Who knows what the thinking and feeling being may be on Saturn? But do feeling and thought exist on Saturn? Why not? Perhaps the feeling and thinking being on Saturn has more senses than I have? If that is so, ah, how wretched is the Saturnian! The more senses, the more needs.

Bordeu: He is right: organs produce needs, and reciprocally, needs produce organs.

Mlle. de l'Espinasse: Doctor, are you raving too?

Bordeu: But why not? I have seen two stumps end by becoming two arms.

Mlle. de l'Espinasse: That's a lie.

Bordeu: True; but, where the two arms were lacking, I have seen the shoulder-blades grow long, move together like pincers, and become two stumps.

Mlle. de l'Espinasse: That's nonsense.

Bordeu: It's a fact. Assume a long succession of armless generations, assume continual efforts, and you will see the two ends of this pincer stretch out, stretch further and further, cross at the back, come round in front, perhaps develop fingers at their ends, and make arms and hands once more. The original conformation degenerates or is perfected by necessity and by normal function. We walk so little, we work so little and we think so much, that I don't despair that man may end by being only a head.

Mlle. de l'Espinasse: A head! A head! That's not very much; I hope that excessive love-making won't... But you're suggesting some very ridiculous ideas to me.

Bordeu: Hush!

d'Alembert: So I am what I am, because I had to be so. Change the whole, and you will necessarily change me; but the whole is constantly changing... man is merely a common product, the monster an uncommon product; both equally natural, equally necessary, equally part of the universal and general order of things. And what is astonishing about that? All creatures intermingle with each other, consequently all species... everything is in perpetual flux. Every animal is more or less man; every mineral is more or less plant; every plant more or less animal. There is nothing precise in nature... Father Castel's ribbon.[9] Yes, Father Castel, it's your ribbon and nothing more. Everything is more or less one thing or another, more or less earth, more or less water, more or less air, more or less fire; everything belongs more or less to one kingdom or another... therefore nothing is of the essence of a particular being. No, surely, since there is no quality of which no being has a share... and that it is the greater or less degree of this quality that makes us attribute it to one being to the exclusion of another. And you talk of individuals, poor philosophers! Stop thinking of individuals; answer me. Is there in nature one atom that strictly resembles another atom? No. Don't you agree that everything is connected in nature, and that it is impossible that there should be a missing link in the chain? Then what do you mean by your individuals? There aren't any, no, there aren't any. There is only one great individual, that is the whole. In that whole, as in a machine or some animal, you may give a certain name to a certain part, but if you call this part of the whole an individual you are making as great a mistake as if you called the wing of a bird, or a feather on that wing, an individual...[10] And you talk of essences, poor philosophers! Leave your essences out of it. Consider the general mass, or if your imagination is too feeble to embrace that, consider your first origin and your latter end. O Architas! you who measured the globe, what are you? A handful of ashes. What is a being? The sum of a certain number of tendencies. Can I be anything other than a tendency? No, I am moving towards an end. And species? Species are only tendencies towards a common end which is peculiar to them. And life? Life, a succession of actions and reactions.[11] Living, I act and react as a mass... dead, I act and react in the form of molecules. Then I do not die? No, no doubt, I don't die in that sense, neither I myself nor anything else. Birth, life, decay, are merely changes of form. And what does the form matter? Each form has the happiness and misfortune which pertain to it. From the elephant to the flea, from the flea to the sensitive living atom, the origin of all, there is no point in nature but suffers and enjoys.

Mlle. de l'Espinasse: He says nothing more.

Bordeu: No. That was a fine flight he made; that was very lofty philosophy: only theoretical at the moment, yet I believe that the more progress is made in human knowledge, the more will its truth be confirmed.

Mlle. de l'Espinasse: And where had we got to meanwhile?

Bordeu: Really, I don't remember; he suggested so many phenomena to my mind while I was listening to him!

Mlle. de l'Espinasse: Wait, wait ... I'd got as far as my spider.

Bordeu: Yes, yes.

Mlle. de l'Espinasse: Come here, Doctor. Imagine a spider in the centre of its web. Shake one thread, and you will see the watchful creature run up. Well! What if the thread that the insect draws out of its intestines, and draws back thither when it pleases, were a sensitive part of itself?

Bordeu: I understand you. You imagine inside yourself, somewhere, in some corner of your head, in that part for instance that is called the *meninges*, one or several points to which are referred back all the sensations aroused along the threads.

Mlle. de l'Espinasse: Exactly.

Bordeu: Your idea is perfectly correct; but don't you see that it comes to much the same thing as a certain cluster of bees?

Mlle. de l'Espinasse: Why, so it does; I've been speaking prose without knowing it.

Bordeu: And very good prose too, as you will see. Anyone who knows man only in the form he appears in at birth, has not the slightest idea what he is really like. His head, his feet, his hands, all his limbs, all his viscera, all his organs, his nose, his eyes, his ears, his heart, his lungs, his intestines, his muscles, his bones, his nerves, his membranes, are, properly speaking, only the gross developments of a network that forms itself, increases, extends, throws out a multitude of imperceptible threads.[12]

Mlle. de l'Espinasse: That's my web; and the point whence all these threads originate is my spider.

Bordeu: Perfect.

Mlle. de l'Espinasse: Where are the threads? Where is the spider placed?

Bordeu: The threads are everywhere; there is no point on the surface of your body which their ends do not reach; and the spider has its seat in the part of your head that I have mentioned, the *meninges,* the slightest touch on which would make the whole organism fall into torpor.

Mlle. de l'Espinasse: But if an atom sets one of the threads of the web quivering, the spider is alarmed and disturbed, runs away or comes hurrying up. At the centre it learns all that is happening in any part of the huge chamber over which it has spun its web. Why can I not know what is happening in my chamber, the world, since I am a group of sensitive points, pressing on everything and subject to impressions from everything?

Bordeu: Because impressions grow weaker in proportion to the distance from which they come.

Mlle. de l'Espinasse: If the lightest blow is struck at the end of a long beam, I hear that blow, if I have my ear placed to the other end. If this beam stood touching the Earth with one end and Sirius with the other, the same effect would be produced. Why, since everything is connected, contiguous, so that this beam exists in reality, do I not hear what is happening in the vast space that surrounds me, especially if I listen attentively?

Bordeu: And who has told you that you don't hear it, more or less? But the distance is so great, the impression is so weak and interrupted by so many others crossing its path; you are surrounded and deafened by such violent and diverse noises; the reason being that, between Saturn and you there are only contiguous bodies, whereas there should be continuity.

Mlle. de l'Espinasse: It's a great pity.

Bordeu: True, for then you would be God. Through your identity with all the beings in nature, you would know all that happens; through your memory, you would know all that has happened.

Mlle. de l'Espinasse: And all that is going to happen?

Bordeu: You would form, about the future, conjectures that were likely but liable to error. It's just as if you sought to guess what is going to happen inside yourself, at the tip of your foot or your hand.

Mlle. de l'Espinasse: And who has told you that this world has not also got its '*meninges,*' that there is not, dwelling in some corner of space, a large or a small spider whose threads reach out to everything?

Bordeu: No one; and still less, whether it has ever existed or ever will exist.

Mlle. de l'Espinasse: Could a God of that sort...

Bordeu: The only sort that is conceivable...

Mlle. de l'Espinasse: ...have existed, or come into existence and pass away?

Bordeu: No doubt; but since he would be a material part of the material universe, subject to vicissitudes, he would grow old and die.

Mlle. de l'Espinasse: But now another extravagant idea comes into my mind.

Bordeu: I'll excuse you from telling it, I know what it is.

Mlle. de l'Espinasse: Well then, what is it?

Bordeu: You picture intelligence combined with highly energetic portions of matter, and the possibility of every imaginable sort of prodigy. Others have thought like you.

Mlle. de l'Espinasse: You have guessed my thought, and I think none the better of you for it. You must have a remarkable tendency towards madness.

Bordeu: Granted. But what is there terrifying about that idea? There would be an epidemic of good and evil geniuses; the most constant laws of nature would be interrupted by natural agents; our physical science would become more difficult thereby, but there wouldn't be any miracles.

Mlle. de l'Espinasse: Truly, one must be very circumspect about what one affirms and what one denies.

Bordeu: To be sure, anyone who described to you a phenomenon of this sort would seem a mighty liar. But let us leave all these imaginary beings, not excepting your spider with its infinite network; let's get back to your own being and its formation.

Mlle. de l'Espinasse: I'm willing.

d'Alembert: Mademoiselle, you are with someone; who is that talking to you?

Mlle. de l'Espinasse: It's the doctor.

d'Alembert: Good morning, Doctor; what are you doing here so early?

Bordeu: You shall hear later: go to sleep now.

d'Alembert: I certainly need to. I do not think I ever passed a more restless night than this one. Don't go away before I am up.

Bordeu: No. I'll wager, Mademoiselle, that you have assumed that you were at twelve years old a woman half your present size, at four years a woman half as small again, as a foetus a tiny woman, in your mother's ovaries a very tiny woman, and that you have always been a woman in the same shape as today, so that only your successive increases in size have made all the difference between yourself at your origin and yourself as you are today.

Mlle. de l'Espinasse: I admit it.

Bordeu:[13] And yet nothing is further from the truth than this idea. At first you were nothing at all. You began as an imperceptible speck, formed from still smaller molecules scattered through the blood and lymph of your father and mother; that speck became a loose thread, then a bundle of threads.[a] Up till then, not the slightest trace of your own agreeable form; your eyes, those fine eyes, were no more like eyes than the tip of an anemone's feeler is like an anemone. Each of the fibres in the bundle of threads was transformed solely by nutrition and according to its conformation, into a particular organ; exception being made of those organs in which the fibres of the bundle arc metamorphosed, and to which they give birth.[b] The bundle is a purely sensitive system;[c] if it continued under that form, it would be susceptible to all those impressions that affect simple sensitivity, such as cold and heat, softness and harshness. These impressions, experienced successively, varied amongst themselves and each varying in intensity, might perhaps produce memory, self-consciousness, a very limited form of reason. But this pure and simple sensitivity, this sense of touch, is differentiated through the organs that arise from each separate fibre;[d] one fibre, forming an ear, gives rise to a kind of touch that we call noise or sound; another forming the palate, gives rise to a second kind of touch that we call taste; a third, forming the nose and its inner lining, gives rise to a third kind of touch that we call smell; a fourth, forming an eye, gives rise to a fourth kind of touch that we call colour.

Mlle. de l'Espinasse: But, if I've understood you aright, those who deny the possibility of a sixth sense, a real hermaphrodite, are very stupid. Who has told them that nature could not form a bundle with a peculiar fibre which would give rise to an organ unknown to us?[e]

Bordeu: Or with the two fibres that characterize the two sexes?[f] You are right; it's a pleasure to talk with you; not only do you follow what is said to you, but you draw from it conclusions that astonish me by their soundness.

Mlle. de l'Espinasse: Doctor, you're saying that to encourage me.

Bordeu: No, on my word, I'm saying what I really think.

Mlle. de l'Espinasse: I can quite well see the purpose of some of the fibres in the bundle; but what becomes of the others?

Bordeu: And do you think any other woman but yourself would have thought of that question?

Mlle. de l'Espinasse: Certainly.

Bordeu: You're not vain. The rest of the fibres[g] go to form as many different kinds of touch as there are different organs and parts of the body.

Mlle. de l'Espinasse: And what are they called? I never heard speak of them.

Bordeu: They have no name.

Mlle. de l'Espinasse: Why not?

Bordeu: Because there is less difference between the sensations excited through their means, than there is between the sensations excited by means of the other organs.

Mlle. de l'Espinasse: In all seriousness, do you believe that the foot, the hand, the thighs, the belly, the stomach, the chest, the lungs, the heart, have their own particular sensations?

Bordeu: I do believe so. If I dared, I would ask you if, among those sensations that are not named.

Mlle. de l'Espinasse: I understand you. No. That one is quite unique of its kind, the more's the pity. But what reason have you for assuming this multiplicity of sensations, more painful than pleasant, which you are pleased to bestow on us?

Bordeu: The reason? That we distinguish them to a considerable extent. If this infinite variety of touch did not exist we should know that we experienced pleasure or pain but we should not know where they arose. We should need the aid of sight. It would no longer be a question of sensation, but of experiment and observation.

Mlle. de l'Espinasse: Then, if I should say my finger hurt, and I were asked why I declared it was my finger that hurt, I should be obliged to say, not that I felt it hurt, but that I felt pain and that I saw my finger was injured.

Bordeu: That's it. Come and let me kiss you.

Mlle. de l'Espinasse: With pleasure.

d'Alembert: Well done, Doctor, you are kissing Mademoiselle.

Bordeu: I have thought over this problem a great deal, and it seems to me that the direction and the place whence the shock arises would not be enough to determine the judgment immediately passed by the centre of the bundle.

Mlle. de l'Espinasse: I don't know about that.

Bordeu: I appreciate your doubt. It is so common to take natural qualities for acquired habits almost as old as ourselves.

Mlle. de l'Espinasse: And reciprocally too.

Bordeu: Be that as it may, you see that in a question that concerns the first formation of the animal, you are starting too late if you observe and consider only the fully formed animal; that you need to go back to its first rudiments, and that it is therefore desirable to strip off your existing organization, and to go back to a moment when you were merely a soft, filamentous, shapeless, worm-like substance, more analogous to the bulb or root of a plant than to an animal.

Mlle. de l'Espinasse: If it were the custom to go naked in the streets I should be neither the first nor the last to conform to it. So, do what you like with me, as long as I learn something. You told me that every fibre in the bundle formed a particular organ; what proof have you that this is so?

Bordeu: Do in your mind what nature sometimes does actually; deprive the bundle of one of its fibres, for instance of the fibre which should form the eyes; what do you think will happen?

Mlle. de l'Espinasse: Perhaps the animal will have no eyes.

Bordeu: Or one single one in the middle of its forehead.

Mlle. de l'Espinasse: It would be a Cyclops.

Bordeu: Yes, a Cyclops.

Mlle. de l'Espinasse: The Cyclops, then, may not be a purely fabulous creature?

Bordeu: So far from it, that I can show you one whenever you like.

Mlle. de l'Espinasse: And who knows the cause of this peculiarity?

Bordeu: The man who has dissected the monster and found that it has only one optic nerve. Do mentally what nature sometimes does actually; suppress the fibre of the bundle which should form the ear, the animal will have no ears, or only one, and the anatomist will find on dissection neither the olfactory nerves nor the auditory nerves, or will find only one of these. Go on suppressing the fibres, and the animal will lack a head, feet, hands; it will last but a short time, but it will have lived.[14]

Mlle. de l'Espinasse: And are there examples of this?

Bordeu: Assuredly. And that's not all. Duplicate some of the fibres of the bundle, and the animal will have two heads, four eyes, four ears, three testicles, three feet, four arms, six fingers on each hand. Disturb the fibres of the bundle, and the organs will be out of place; the head will be in the middle of the chest, the lungs will be on the left the heart on the right. Stick two fibres together, and the organs will be fused together; the arms will cling to the body, the thighs, legs and feet will be joined up, and you will have every conceivable sort of monster.

Mlle. de l'Espinasse: But it seems to me that so complex a system as an animal, an organism which is born from a speck, from a seething fluid, perhaps from two fluids mingled haphazard, since one hardly knows what one's doing on these occasions; an organism which advances towards perfection by an infinite number of successive developments; an organism the regular or irregular structure of which depends on a bundle of thin, loose, flexible fibres, a sort of skein in which the slightest fibre cannot be broken, snapped, displaced or removed without distressing consequences for the whole, such an organism should become even more frequently tangled up in the place of its formation than do my silks on my bobbin.[14]

Bordeu: And in fact, the organism does suffer much more than people think. There is not enough dissection done, and ideas about its formation are very far from the truth.

Mlle. de l'Espinasse: Are there striking examples of these peculiar deformities at origin, other than hunchbacks and cripples, whose misshapen state might be attributed to some hereditary defect?

Bordeu: There are countless examples, and quite recently there died at the hospital of la Charite in Paris, at the age of twenty-five, following an inflammation of the lungs, a carpenter called Jean-Baptiste Mace, native of Troyes, who had the internal viscera of the chest and the abdomen transposed, the heart on the right, whereas you have it on the left; the liver on the left; the stomach, the spleen, the pancreas on the right hypochondrium; the *vena porta* to the liver on the left side, instead as it should be to the liver on the right; a similar transposition of the alimentary canal; the kidneys, back to back against the *vertebrae* or the loins, were in the shape of a horseshoe. And now let them talk about final causes!

Mlle. de l'Espinasse: It's very odd.

Bordeu: If Jean-Baptiste Mace had married and had children...

Mlle. de l'Espinasse: Well, Doctor, these children?

Bordeu: Would be formed in the normal way; but some one of their children's children, after a hundred years or so, since these irregularities make leaps, will revert to the extraordinary conformation of his ancestor.

Mlle. de l'Espinasse: And what causes these leaps?

Bordeu: Who knows? It takes two to make a child, as you know. It may be that one of the agents counteracts the other's defect, and that the faulty network only reappears when the descendant of the monstrous breed is dominant and controls the formation of the network. The bundle of fibres constitutes the original primary difference between all species of animals. The varieties in the form of the bundle of each species constitute the monstrous varieties within that species.[15]

(*After a long silence, Mlle. de l'Espinasse emerged from her reverie and awoke the doctor from his by the following question*):

Mlle. de l'Espinasse: I have just had a very mad idea.

Bordeu: What's that?

Mlle. de l'Espinasse: Man may be merely a monstrous form of woman, or woman a monstrous form of man.

Bordeu: You would have had that idea much sooner, if you had known that a woman has all a man's organs, and that the only difference between them is that between a bag hanging down outside, and an inverted bag inside; that a female foetus looks deceptively like a male foetus; that the part that causes this confusion is gradually effaced in the female foetus, as the interior bag grows bigger; that it is never obliterated to the point of losing its original form, but keeps this form on a small scale; that it is liable to the same movements, that it, too, gives rise to the voluptuous impulse; that it has its *glans*, its foreskin, and that on the tip of it there can be seen a point which appears to be the opening of a urinary canal that is now closed; that there is in man, from the *anus* to the *scrotum*, a space called the *perineum*, and from the *scrotum* to the tip of the *penis*, a scar that looks like a sewn-up *vulva*; that women whose *clitoris* is over-developed grow beards; that eunuchs are beardless, while their thighs broaden, their hips curve, their knees grow rounded, and that, by losing the characteristic organization of one sex, they seem to revert to the characteristic conformation of the other. Those Arabs who have become castrated through continual horseback-riding lose their beards, develop a high voice, dress like women, ride with the women in the wagons, squat to urinate, and assume female ways and customs. But we have wandered far from our objective. Let us get back to our bundle of animated and living filaments.

d'Alembert: I think you are talking filth to Mlle. de l'Espinasse.

Bordeu: When one talks about science one has to use technical terms.

d'Alembert: You are right; then they lose the train of associated ideas that would make them indecent. Go on, Doctor. You were saying to Mademoiselle that the womb is only a *scrotum* turned inside out, during which process the ovaries were ejected from the bag that contained them and thrown right and left in the cavity of the body; that the *clitoris* is a tiny male member; that this woman's member gets gradually smaller as the womb or inverted *scrotum* grows longer, and that...

Mlle. de l'Espinasse: Yes, yes, be quiet and don't interrupt us.

Bordeu: You see, Mademoiselle, that, when we examine our sensations in general, which are all merely a differentiated sense of touch, we must neglect the successive forms assumed by the network, and consider only the network itself.

Mlle. de l'Espinasse: Every filament of the sensitive network can be hurt or stimulated along its whole length. Pleasure or pain is here or there, in one spot or another along the prolonged legs of my spider, for I always come back to my spider; that spider is the common origin of all the legs and their prolongations, and refers the pain or the pleasure to such and such a place without feeling it.

Bordeu: It is the constant and unvarying communication of all impressions to this common origin which constitutes the unity of the animal.

Mlle. de l'Espinasse: It is the recollection of all these successive impressions which makes up, for each animal, the story of its life and of its individual being.

Bordeu: While memory, and the process of comparison, which inevitably result from all these impressions, form thought and reasoning power.

Mlle. de l'Espinasse: And where does this process of comparison take place?

Bordeu: At the origin of the network.

Mlle. de l'Espinasse: And this network? ...

Bordeu: Has, at its origin, no sense peculiarly its own; it does not see, hear, or suffer. It is produced and nourished; it emanates from a soft, insensitive, inert substance, that serves it as a pillow, seated on which it listens, judges and decides.

Mlle. de l'Espinasse: It feels no pain?

Bordeu: No; the slightest pressure cuts short its power to judge and the whole animal falls into a death-like condition. Remove the pressure, and the judge resumes its functions, and the animal lives again.

Mlle. de l'Espinasse: And how do you know this? Has a man ever been made to die and live again at will?

Bordeu: Yes.

Mlle. de l'Espinasse: And how was that?

Bordeu: I will tell you; it is a curious fact. La Peyronie, whom you may have known, was summoned to a patient who had received a violent blow on the head. This patient felt a throbbing there. The surgeon had no doubt that an abscess had formed in the brain, and that there was not a moment to lose. He shaved the patient's head and trepanned him. The point of the instrument fell exactly in the centre of the abscess. The pus was formed; he emptied it out; he cleaned the abscess with a syringe. When he drove the injection into the abscess, the sick man closed his eyes; his limbs remained inactive, motionless, without the slightest sign of life; when the injection was pumped out again, and the origin of the bundle relieved of the weight and pressure of the injected fluid, the sick man opened his eyes again, moved, spoke, felt, was reborn and lived.

Mlle. de l'Espinasse: That is very odd; and did the patient recover?

Bordeu: He recovered; and when he was well, he could reflect, think, reason, he had the same wit, the same good sense, the same acuteness, though lacking a considerable portion of his brain.

Mlle. de l'Espinasse: This judge of yours is a most extraordinary creature.

Bordeu: He, too, makes mistakes at times; he is subject to errors due to habit; one feels pain in a limb which one no longer has. You can deceive him when you wish; cross two of your fingers over each other, touch a little ball, and the judge will declare that there are two.

Mlle. de l'Espinasse: That's because he is like all the judges in the world, and needs experience, without which he would mistake the feeling of ice for that of fire.

Bordeu: He goes further than that; he may attribute an almost infinite volume to an individual, or else concentrate him almost to a point.

Mlle. de l'Espinasse: I don't understand.

Bordeu: What limits your real extension, the true sphere of your faculty of sensation?

Mlle. de l'Espinasse: My sight and my sense of touch.

Bordeu: By day; but what limits it at night, in darkness, especially when you are thinking of something abstract, and even by day, when your mind is preoccupied?

Mlle. de l'Espinasse: Nothing does. I exist as it were within a single point; I almost cease to be material, I feel nothing but my thought: I am no longer conscious of place or movement, body, distance or space: the universe is abolished for me, and I am as nothing to it.

Bordeu: That is the final term in the concentration of your being; but its imaginary expansion can be limitless. When the true limit of your sensitiveness is exceeded, either by condensing yourself within yourself or by extending beyond yourself, there is no knowing what may result.

Mlle. de l'Espinasse: Doctor, you are right. It has often seemed to me in dreams...

Bordeu: And to sick people during an attack of gout.

Mlle. de l'Espinasse: That I was becoming vast...

Bordeu: That their feet touched the canopy of their bed.

Mlle. de l'Espinasse: That my arms and legs were stretching out to infinity, that the rest of my body was growing in proportion; that the Enceladus of legend was a pigmy to me, that Ovid's Amphitriton, whose long arms made a huge girdle round the Earth, was but a dwarf by my side, and that I scaled the heavens and embraced the two hemispheres.

Bordeu: Very fine. And I have known a woman who experienced the same phenomenon in the opposite sense.

Mlle. de l'Espinasse: What! Did she grow smaller by degrees and shrink within herself?

Bordeu: To the point of feeling herself as thin as a needle; she could see, hear, reason and judge; she was in mortal fear of losing herself, shuddered at the approach of the smallest objects and scarcely dared move from her place.

Mlle. de l'Espinasse: That is a peculiar dream, most unpleasant and inconvenient.

Bordeu: It was no dream, but one of the symptoms accompanying the cessation of the menstrual flow.

Mlle. de l'Espinasse: And did she remain long in the shape of a tiny imperceptible woman?

Bordeu: For an hour or so, after which she would gradually regain her normal volume.

Mlle. de l'Espinasse: And what is the reason for these queer sensations?

Bordeu: In their natural and quiet state, the fibres that make up the bundle have a certain degree of tension; a customary tone and energy that limits the extent – real or imagined – of one's body. I say real or imagined, for this tension, this tone, this energy being variable, our body has not always the same volume.

Mlle. de l'Espinasse: Then, physically as well as morally, we are liable to fancy ourselves greater than we are?

Bordeu: Cold makes us shrink, heat makes us expand, and an individual may go through life thinking himself smaller or bigger than he really is. If the bulk of the bundle should happen to undergo a violent irritation – if the fibres stand erect and their innumerable tips suddenly stretch out beyond their accustomed limits, then the head, the feet, the other members, every point over the surface of the body will be projected to an immense distance, and the individual will feel himself a giant. The contrary phenomenon will take place if a gradual insensitiveness, apathy and inertia take hold of the tips of the fibres and creep gradually towards the origin of the bundle.

Mlle. de l'Espinasse: I can imagine that such expansion could never be measured, and I can also imagine that this insensitiveness, apathy and inertia of the tips of the fibres, this numbness, having progressed a certain distance, might be checked and halted.

Bordeu: As happened to La Condamine; then the person feels as if he had balloons under his feet.[16]

Mlle. de l'Espinasse: He exists beyond the limits of his sensitiveness, and if this apathy were to enfold him in every direction, he would appear as a tiny man living within a dead man.

Bordeu: From this you may conclude that the animal which was to begin with a mere point, does not yet know whether he is anything more than that. But let us get back.

Mlle. de l'Espinasse: To what?

Bordeu: To La Peyronie's trepanning. I fancy you have there what you asked for, an instance of a man living and dying alternately. But there is a better one.

Mlle. de l'Espinasse: And what may that be?

Bordeu: The fable of Castor and Pollux in real life; two children, in whose case the life of one was immediately followed by the death of the other, and the life of the latter immediately followed by the death of the first.

Mlle. de l'Espinasse: Oh, that's a tall story. And how long did this go on?

Bordeu: This existence lasted for two days, which they shared equally and alternately, so that each had for its portion one day of life and one of death.

Mlle. de l'Espinasse: I'm afraid, Doctor, that you are taking advantage of my credulity. Take care, for if you deceive me once I shall never trust you again.

Bordeu: Do you ever read the *Gazette de France*?

Mlle. de l'Espinasse: Never, although it is the masterpiece of two clever men.

Bordeu: Borrow the issue of the fourth of this month, September, and you will see that at Rabastens, in the diocese of Albi, two girls were born back to back, joined by their lowest lumbar *vertebra*, their buttocks and the lower part of the trunk. One could not be held upright without the other's head being upside down. When laid down they were face to face; their thighs were bent between their trunks, their legs in the air; in the centre of the common circular line that connected them through their lower abdomens, the sexual organs could be discerned, and between the right thigh of one which corresponded to the left thigh of her sister, there was, in a hollow, a little *anus* through which the *meconium* flowed out.

Mlle. de l'Espinasse: What a peculiar species of creature!

Bordeu: They took some milk which was given them in a spoon. They lived for twelve hours as I have told you, one losing consciousness as the other regained it, one dying while the other lived. The first swoon of one and the first life of the other lasted four hours, the subsequent alternating swoons and returns to life were shorter; they expired at the same instant. It was observed that their navels went in and stood out alternately; that of the child who was unconscious was sucked in, while that of the child who was coming back to life stood out.

Mlle. de l'Espinasse: And what can you say about these alternations of life and death?

Bordeu: Nothing significant perhaps; but as one sees everything through the spectacles of one's pet theory, and I don't want to be an exception to that rule, I say, that it is the same phenomenon as that of the trepanned patient of La Peyronie's, duplicated in two beings joined together; that the networks of these children were so thoroughly interconnected that they acted and reacted on one another; when the origin of the network of one of them predominated, it affected the network of the other, who immediately lost consciousness; the reverse happened if the latter's network were dominant in the common system. In the case of La Peyronie's patient, the pressure was from above downwards through the weight of a fluid; in the case of the twins of Rabastens, it was from below upwards, through traction of a certain number of the fibres of the network; a conjecture which is borne out by the alternating movements of the two navels, a movement outwards in the child that was reviving, a movement inwards in the one which was dying.

Mlle. de l'Espinasse: And there we have two souls linked together.

Bordeu: One animal with the rudiments of twofold senses and twofold consciousness.

Mlle. de l'Espinasse: And yet able to enjoy only one at a time; but who knows what might have happened if that animal had lived?

Bordeu: What sort of communication might have been set up between the two brains, by the common experience of every instant of life, the strongest bond of habit imaginable?

Mlle. de l'Espinasse: Double senses, a double memory, a double imagination, a double power of concentration, one half of a creature observing, reading, meditating while its other half is resting; the latter taking up the same functions when its companion is weary, life doubled for a double being.

Bordeu: It is possible; and since nature, in the course of time, brings about all that is possible, she will form some such strange composite being.

Mlle. de l'Espinasse: What poor creatures we should be, compared with such a being!

Bordeu: But why? A single mind is subject to so many uncertainties, contradictions and absurdities that I cannot imagine what a double mind might not produce. But it is half-past ten, and I can hear a patient calling me all the way from the outskirts of the town.

Mlle. de l'Espinasse: Would he be in great danger if you did not see him?

Bordeu: Less, perhaps, than if I did. If nature does not do her business without me, we shall find it hard to do it together, and I shall certainly not succeed without her.

Mlle. de l'Espinasse: Stay here then.

d'Alembert: Doctor, one word more and then I send you to your patient. Through all the changes I have undergone in the course of my existence, perhaps not having now a single one of the molecules which formed me at birth, how have I maintained my identity for others and for myself?

Bordeu: You told us yourself in your dream.

d'Alembert: Have I been dreaming?

Mlle. de l'Espinasse: All night long, and it sounded so like delirium that I sent for the doctor this morning.

d'Alembert: And all because a certain spider's legs were moving of their own accord, kept the spider on the watch, and made the animal talk. And what did the animal say?

Bordeu: That it was through memory that he maintained his identity for others and for himself; and, let me add through the slowness of the changes. If you had passed in the twinkling of an eye from youth to decay, you would have been thrown into the world as at the first moment of birth; you would not have been yourself in your own eyes, nor in those of others; while they would not have been themselves in your eyes. All connecting links would have been destroyed; all that makes up the history of your life for me, all that makes up the history of my life for you, thrown into confusion. How could you have known that this man, leaning on a stick, his eyes grown dim, dragging himself along with difficulty, and even more unlike himself inwardly than outwardly, was the same who, the day before, walked so lightly, lifted heavy burdens, gave himself up to the deepest meditations, the pleasantest and the most strenuous forms of exercise? You would not have understood your own works, you would not have recognized yourself nor any one else, and no one would have recognized you; all the world's scene would have changed. Consider that there was less difference between yourself at birth and yourself in youth, than there would be between yourself as a young man and yourself grown suddenly decrepit. Consider that, although your birth was linked to your youth by an unbroken series of sensations, yet the first three years of your life form no part of your life-story. Then what would the days of your youth have meant to you if nothing linked them to the period of your decay? D'Alembert grown old would not have the slightest recollection of d'Alembert young.

Mlle. de l'Espinasse: In the cluster of bees, not one would have had time to take on the spirit of the whole.

d'Alembert: What's that you're saying?

Mlle. de l'Espinasse: I am saying that the monastic *spirit* is preserved, because the monastery repeoples itself gradually, and when a new monk enters it he finds a hundred old ones, who induce him to think and feel as they do. When one bee goes, its place in the cluster is taken by another that rapidly adapts itself.

d'Alembert: Come, you are crazy with your talk of monks, bees, clusters and convents.

Bordeu: Not as crazy as you might think. Although the animal has only one consciousness, it has an infinite number of wills; each organ has its own.

d'Alembert: What do you mean by that?

Bordeu: I mean that the stomach desires food, while the palate will have none of it; that the difference between the whole animal on one hand and the stomach and palate on the other is that the animal knows what it wants, while the stomach and palate want without knowing it; and the palate and the stomach are related like man to brute. The bees lose individual consciousness and retain their appetites and wills. The fibre is a simple animal, man a complex animal; but we will keep this text for another time. It does not take so great an event as decay to take away selfconsciousness from man. A dying man receives the sacraments with the deepest piety, confesses his sins, asks forgiveness of his wife, embraces his children, summons his friends, speaks to his physician, gives orders to his servants; he dictates his last wishes, sets his affairs in order, and all this with complete sanity and presence of mind; he recovers, he is convalescent, and he has not the slightest idea of what he has said and done during his illness. That interval, though sometimes a very long one, has disappeared from his life. There are even instances of persons resuming the conversations or the actions which the sudden attack of illness had interrupted.

d'Alembert: I remember that during a public debate, a college pedant, inflated with learning, was beaten by a Capuchin monk whom he had despised. He, beaten, and by whom? By a Capuchin! And on what topic? On the contingent future! on that science of cause and effect which he had studied all his life! And in what circumstances? Before a numerous assembly! Before his pupils! Behold him disgraced. He worried his head over these things so much that he fell into a lethargy that deprived him of all the learning he had acquired.

Mlle. de l'Espinasse: But that was a blessing.

d'Alembert: Why, yes, you're right. He kept his natural senses, but he forgot everything. He was taught afresh to speak and read, and he died just as he was beginning to spell tolerably well. This man was not devoid of gifts; it seems, even, that he had a certain eloquence.

Mlle. de l'Espinasse: Since the doctor has heard your story, he must hear mine too. A young man of eighteen or twenty whose name I forget…

Bordeu: He was a M. de Schullemberg, of Winterthur; he was only fifteen or sixteen.

Mlle. de l'Espinasse: This young man had a fall, and suffered a violent shock to his head.

Bordeu: What do you call a violent shock? He fell from the top of a barn; his skull was fractured, and he remained unconscious for six weeks.

Mlle. de l'Espinasse: Be that as it may, do you know what was the sequel to this accident? The same as in your pedant's case; he forgot all he knew; he went back to his infancy; he had a second childhood, and one which lasted. He was timid and cowardly; he played with toys. If he had been naughty and was scolded, he would go and hide in a corner; he asked leave to pay a big or little 'visit.' He was taught to read and write; but I was forgetting to tell you that he had to learn to walk again. He became a man once more, and a clever man, and he has left a work on natural history.

Bordeu: It is a set of engravings, the plates for M. Zulyer's studies of insects according to the system of Linnæus. I knew about this already: it occurred in the canton of Zurich in Switzerland, and there are many more instances like it. Disturb the origin of the bundle and you change the whole animal; it seems as if it existed there in its entirety, now dominating the branches, now dominated by them.

Mlle. de l'Espinasse: And the animal is either under a despot's rule or under anarchy.

Bordeu: A despot's rule is an apt description. The origin of the bundle commands and all the rest obeys. The animal is master of itself, *compos mentis.*

Mlle. de l'Espinasse: Under anarchy, when all the fibres of the network rise up against their ruler, and there is no longer any supreme authority.

Bordeu: Exactly. In strong fits of passion, in delirium, at times of imminent peril, if the master brings all his subjects' strength to bear in one direction, the weakest animal may display an incredible strength.

Mlle. de l'Espinasse: In the vapours, that variety of anarchy to which we women are peculiarly liable.

Bordeu: There you have the picture of a weak administration, in which everyone claims the supreme authority himself. I know only one way of recovering; it is difficult, but infallible; it is for the origin of the sensitive network, that part that constitutes the individual's identity, to have some powerful motive for regaining its authority.

Mlle. de l'Espinasse: And what happens then?

Bordeu: It happens that it does indeed regain it, or else that the animal perishes. If I had time, I would tell you two curious facts in this connection.

Mlle. de l'Espinasse: But, Doctor, the time of your visit is past, and your patient doesn't expect you any longer.

Bordeu: One should only come here when one has nothing to do, for it's impossible to get away.

Mlle. de l'Espinasse: That burst of ill-temper is quite flattering! But your stories?

Bordeu: For today you'll have to be content with this one. A woman fell into the most alarming hysterical condition, following her confinement; she was subject to uncontrolled fits of weeping and laughing, to chokings, convulsions, heavings of the bosom, gloomy silence, shrill cries, all the most frightful things. This went on for several years. She was passionately in love, and she thought she saw that her lover, weary of her illness, was beginning to drift away from her; then she resolved to be cured or to die. A kind of civil war took place within her, in which now the master had the upper hand, now the subjects. If it happened that the action of the fibres of the network equalled the reaction of their origin, she would fall in a death-like trance; she had to be carried on to her bed and would remain there for hours, motionless and almost lifeless; at other times she suffered only lassitude, general weakness, an exhaustion that looked like being fatal. She persisted in this state of conflict for six months. The rebellion always began in the fibres of the network; she would feel it coming. At the first symptom she would get up, run about, undertake the most violent forms of exercise; she would run up and down the stairs, saw wood, dig the earth. The organ of her will, the origin of the bundle, stiffened its resistance; she said to herself: victory or death. After an infinite number of triumphs and defeats, the ruler maintained the mastery, and the subjects became so submissive that, although this woman has experienced all sorts of domestic troubles and suffered various illnesses, no sign of the hysteria has reappeared.

Mlle. de l'Espinasse: She was brave, but I think I'd have done as much.

Bordeu: Because if you loved at all you would love deeply, and because you are strong.

Mlle. de l'Espinasse: I see. One is strong, if, through habit or through one's organization, the origin of the bundle dominates the fibres; weak if, on the contrary, it is dominated by them.

Bordeu: There are many other conclusions to be drawn from this.

Mlle. de l'Espinasse: But tell us your other story, and you may draw them afterwards.

Bordeu: A young woman had been rather a wanton in her conduct. One day she resolved to put pleasure from her. Living alone, she became subject to melancholy and nervous depression. She sent for me. I advised her to dress like a peasant, to dig the earth all day long, sleep on straw, live on coarse bread. This way of life did not attract her. Then travel, I said. She went all round Europe, and regained health on the high road.

Mlle. de l'Espinasse: That's not what you should have said! But never mind, let's hear your conclusions.

Bordeu: There would be no end to them.

Mlle. de l'Espinasse: All the better: say on.

Bordeu: I haven't the courage.

Mlle. de l'Espinasse: Why not?

Bordeu: Because, at the present rate, we skim the surface of everything and go into nothing deeply.

Mlle. de l'Espinasse: What does that matter? We are only chatting, we are not composing a thesis.

Bordeu: For instance, if the origin of the bundle summons all the strength of the whole to itself, if the entire system is, so to speak, moved in reverse, as I think happens to a man sunk in deep thought, to a fanatic who sees the heavens opened, to the savage who sings in the midst of flames, or in ecstatic trances, in voluntary or involuntary madness.

Mlle. de l'Espinasse: Well?

Bordeu: Well, the animal becomes immune to feeling, exists in a single point. I have not seen that priest of Calamus, spoken of by St. Augustine, who could abstract himself to the point of not feeling burning coals, I have not seen those savages who at the stake, laugh at their enemies, insult them and suggest for themselves more exquisite torments than those they are already suffering; I have not seen in the arena those gladiators who, as they died, remembered the graceful attitudes they had learnt in the gymnasium; but I believe all the facts, because I have seen, seen with my own eyes, an effort as extraordinary as any of these.

Mlle. de l'Espinasse: Doctor, tell it me. I am like children, I love marvellous stories, and when they are to the credit of the human race, I rarely question their truth.

Bordeu: There was in a small town in Champagne, Langres, a good curé called le or de Moni, steeped and imbued with the truth of religion. He had an attack of the stone, and had to be operated on. The day was fixed, the surgeon, his assistants, and myself went to his home; he greeted us with serenity, undressed, lay down; he would not allow himself to be strapped down; 'just put me in the right position'; this was done. Then he asked for a great crucifix which stood at the foot of his bed; it was given him, he clasped it in his arms, he pressed his lips to it. The operation was performed, he did not stir, uttered neither tears nor sighs, and was delivered of his stone without knowing it.

Mlle. de l'Espinasse: That is fine: and after that how can one doubt that he whose breast-bones were shattered with stones, saw the heavens open?

Bordeu: Do you know what earache is like?

Mlle. de l'Espinasse: No.

Bordeu: So much the better for you; it is the cruellest pain of all.

Mlle. de l'Espinasse: Worse than toothache, which I do know, unfortunately?

Bordeu: Incomparably worse. A philosopher, one of your friends, had been tortured by it for a fortnight, and one morning he said to his wife: 'I haven't the courage to get through the day.' He thought that his only hope was to cheat the pain by artifice. Gradually he sunk himself so deep in some problem of metaphysics or geometry, that he forgot his ear. His food was served him, he ate without noticing it; he reached his bedtime without having suffered. The horrible pain only seized him again when the intellectual conflict had ceased, but then it was with an unheard-of ferocity, either because weariness had actually aggravated the complaint or because weakness rendered it less bearable.

Mlle. de l'Espinasse: On emerging from such a condition one must indeed be exhausted: with fatigue; that is what happens sometimes to that man yonder.

Bordeu: It is dangerous, he should take care.

Mlle. de l'Espinasse: I am forever telling him so, but he pays no heed.

Bordeu: The thing is beyond his control now, it has become his life, he will die of it.

Mlle. de l'Espinasse: That sentence frightens me.

Bordeu: What does this exhaustion, this weariness, prove? That the fibres of the bundle have not lain idle and that throughout the whole system there was a violent tension towards a common centre.

Mlle. de l'Espinasse: And if this violent tension, this tendency, should persist, if it should become habitual?

Bordeu: Then you have a nervous habit of the centre of the bundle; the animal is mad, and almost hopelessly so.

Mlle. de l'Espinasse: And why?

Bordeu: A nervous habit of the origin is not like a nervous habit of one of the fibres. The head can command the feet, but the feet cannot command the head; the origin can command one of the fibres, but one of the fibres cannot command the origin.

Mlle. de l'Espinasse: And what is the difference, if you please? Indeed, why cannot the whole of me think? That's a question I should have thought of earlier.

Bordeu: Because consciousness resides only in one place.

Mlle. de l'Espinasse: That's easily said.

Bordeu: Because it can only reside in one place, in the common centre of all sensations, where memory is, where the process of comparison goes on. Each fibre responds only to a certain definite number of impressions, which follow one another separately, unconnected by memory. The origin responds to them all, registers them, retains the recollection or continuous sensation of them, and the animal, from the first moment it is formed, is forcibly led to refer itself thereto, to be concentrated there in its entirety, to exist there.

Mlle. de l'Espinasse: And if my finger could remember?

Bordeu: Your finger would think.

Mlle. de l'Espinasse: And what is memory then?

Bordeu: The property peculiar to the origin of the network, its specific property, just as sight is the property of the eye; and it is no more surprising that memory does not dwell in the eye, than that sight does not dwell in the ear.

Mlle. de l'Espinasse: Doctor, you are evading rather than answering my questions.

Bordeu: I evade nothing, I tell you what I know, and I should know more if the organization of the origin of the network were as familiar to me as that of its fibres, if I had had the same opportunity of observing it. But, if I am weak about particular phenomena, I make up for it where general phenomena are concerned.

Mlle. de l'Espinasse: Such as …?

Bordeu: Reason, judgment, imagination, madness, idiocy, ferocity, instinct.

Mlle. de l'Espinasse: I understand. All these qualities are only consequences of the relation, original or acquired by habit, between the origin of the bundle and its branches.

Bordeu: Exactly. Where the origin or trunk is too vigorous in relation to the branches, you have poets, artists, imaginative people, cowards, lunatics, madmen. Where it is too weak you get so-called brutes and savage beasts. Where the whole system is slack and soft, without energy, you get imbeciles; where the whole system is energetic, harmonious, well-disciplined, you have sound thinkers, philosophers, sages.

Mlle. de l'Espinasse: And according to which branch is dominant, we have the different forms of instinct in animals and the different forms of genius in man; the dog has its scent, the fish its hearing, the eagle its sight; d'Alembert is a geometrician, Vaucanson a mechanical engineer, Grétry a musician, Voltaire a poet; the varied effects of some one fibre in the bundle being stronger in them than any other, and stronger than the corresponding fibre in other beings of the same species.

Bordeu: And there is the tyranny of habit; old men go on loving women, Voltaire goes on writing tragedies.

(*Here the doctor began to muse and Mlle. de l'Espinasse said to him*):

Mlle. de l'Espinasse: Doctor, you are day-dreaming.

Bordeu: True.

Mlle. de l'Espinasse: What about?

Bordeu: Voltaire.

Mlle. de l'Espinasse: Well?

Bordeu: I was wondering what makes a great man.

Mlle. de l'Espinasse: And what is it?

Bordeu: How sensibility...[17]

Mlle. de l'Espinasse: Sensibility?

Bordeu: Or the extreme mobility of certain fibres of the network, is the dominant quality of second-rate people.

Mlle. de l'Espinasse: Oh! Doctor, what blasphemy!

Bordeu: I was expecting that. But what is a being possessed of sensibility? One abandoned to the mercy of his diaphragm; should a pathetic phrase strike his ear, a strange phenomenon meet his eye, of a sudden an inward tumult is set up, all the fibres of the bundle are agitated, a shudder runs through his frame, he is seized with horror, his tears flow, sighs choke him, his voice breaks, and the origin of the bundle does not know what it is doing: farewell to self-control, reason, judgment, instinct and resourcefulness

Mlle. de l'Espinasse: I recognize myself.

Bordeu: The great man, if he has been unlucky enough to receive such a disposition from nature, will ceaselessly strive to weaken it, to dominate it, to gain the mastery over his movements and to let the origin of the bundle retain all the power. Then he will have self-control in the midst of the greatest dangers, he will judge coldly, but sanely. Nothing that might further his desires, help towards his object, will escape him; he will not be easily surprised; at forty-five he will be a great king, a great minister, a great politician, a great artist, above all a great actor, a great philosopher, a great poet, a great musician, a great doctor; he will rule over himself and all around him. He will have no fear of death, that fear which, in the Stoic's sublime phrase, the strong man grasps as a handle to lead the weak man where he wishes; he will have broken that handle and will, at the same time, be delivered from every tyranny in the world. Men of sensibility and madmen are on the stage, he is in the stalls, he is the wise man.

Mlle. de l'Espinasse: God preserve me from the society of such a wise man!

Bordeu: It is for want of striving to be like him that you will experience violent griefs and joys in turn, that you will pass your whole life in laughter and tears, and never grow out of your childhood.

Mlle. de l'Espinasse: I'm resigned to that.

Bordeu: And do you hope it will make you happier?

Mlle. de l'Espinasse: I cannot tell.

Bordeu: Mademoiselle, this quality, that is prized so highly, that leads to nothing great, almost always brings pain when exerted strongly, tedium when exerted mildly; either one is bored or one is intoxicated. You yield yourself without restraint to the enjoyment of some delicious music, you let yourself be carried away by the charm of a pathetic scene; you feel a tightening of the throat, the pleasure passes, and you retain only the sense of suffocation that persists all the evening.

Mlle. de l'Espinasse: But supposing I can enjoy the sublime music or the touching scenes only on these conditions?

Bordeu: You are mistaken. I too can enjoy, I can admire, and I never suffer pain except from colic; my pleasure is pure; my criticism is the more severe thereby, my praise more precious and more deliberate. Is there such a thing as a bad tragedy for souls as easily moved as yours? How often have you not blushed, on reading a play, to think of the ecstasy you experienced at the performance of it, and *vice versa.*

Mlle. de l'Espinasse: Yes, it has happened to me.

Bordeu: Therefore it is not sentimentalists such as you, it is calm, cold persons like myself that have a right to say: 'This is true, this is good, this is beautiful.' Let us strengthen the origin of the network, that is the best thing we can possibly do. Do you know that life depends upon it?

Mlle. de l'Espinasse: Life? That's a serious matter, Doctor.

Bordeu: Yes, life. There is no one who has not, at some time or other, felt sick of living. A single incident may be enough to turn this feeling into an unconscious habit; and then in spite of distractions, of varied amusements, of friends' advice and of one's own efforts, the fibres of the bundle persist in shaking the origin with fatal blows; the wretched victim struggles in vain, the whole scene of the universe grows dark for him; he walks escorted by a relentless band of gloomy thoughts, and ends by casting off the burden of himself.

Mlle. de l'Espinasse: Doctor, you frighten me.

d'Alembert (who has got up and is wearing a dressing-gown and night-cap): And sleep, Doctor, what have you to say about that? Sleep is a good thing.

Bordeu: Sleep, that state in which, either through exhaustion or through habit, the whole network slackens and stays motionless, where, as in sickness, each separate fibre of the network stirs, quivers, sends back to the common origin a swarm of sensations, often incongruous, disconnected, confused; at other times so linked up, so consistent, so well-ordered that a waking man could not be more reasonable, more eloquent, more imaginative; sometimes so powerful and so vivid that one remains in doubt, on waking, whether the thing didn't really happen. ...

Mlle. de l'Espinasse: Well, what about sleep?

Bordeu: It is a condition of the animal in which there is no more unity; all harmony, all discipline ceases. The master is left to the mercy of the subjects, and the unbridled energy of his own activity. Should the optic fibre quiver, the origin of the network sees; should the auditory fibre urge it, it hears. Action and reaction alone subsist between them; which follows from the property of the centre, from the law of continuity and from habit. Should action begin by the voluptuous fibre, destined by nature for the pleasures of love and the propagation of the species, the effect of the reaction at the origin of the bundle will be to call up the image of the loved one. If, on the contrary, this image is first called up at the origin of the bundle, the result of the reaction will be a tension of the voluptuous fibre, effervescence and effusion of the seminal fluid.

d'Alembert: So there is an upward dream and a downward dream. I had one of those last night; but which direction it took I couldn't say.

Bordeu: When one is awake, the network responds to the impressions of external objects. In sleep, all that it experiences springs from the exercise of its own sensitiveness. In dreams there is no distraction, hence their vividness;[18] they nearly always result from some irritation, some temporary disorder. The origin of the network is alternately active and passive, in an infinite number of ways; hence its confusion. Sometimes, in dreams, concepts are as connected and distinct as when the animal is in direct contact with the natural scene. It is simply that the image of this scene has been called up afresh; hence the realism of the dream, hence the impossibility of distinguishing it from the waking state; there is no greater probability in favour of one of these states than of the other; experiment alone will indicate the error.

Mlle. de l'Espinasse: And is experiment always possible?

Bordeu: No.

Mlle. de l'Espinasse: If, in a dream, I see a friend that I have lost, see him as vividly as though he were still in existence; if he speaks to me and I hear him; if I touch him and he seems solid to my hands; if on waking I feel my heart full of tender emotion and grief and my eyes overflowing with tears; if my arms are still outstretched towards the spot where he appeared to me, what will convince me that I have not really seen, heard and touched him?

Bordeu: The fact of his absence. But, if it is impossible to distinguish sleep from the waking state, who can judge its duration? A quiet sleep is an unconscious interval between bed-time and rising-time; a troubled sleep may seem to last for years. In the first case, at any rate, consciousness of one's identity ceases entirely. Can you tell me of one dream that has never been dreamt and never will be?

Mlle. de l'Espinasse: Yes, to dream that one is somebody else.

d'Alembert: And in the second case, one is not only conscious of one's identity but also of one's will and of one's liberty. What are the will and liberty of a dreaming man?

Bordeu: What are they? The same as those of a waking man; the latest impulse of desire and aversion, the last result of all that one has been from birth to the actual moment; and I defy the subtlest mind to perceive the least difference between them.

d'Alembert: Do you think so?

Bordeu: And it's you who ask me that! You, who, absorbed in profound speculations, have passed two-thirds of your life dreaming with your eyes open and doing involuntary actions; yes, far more involuntary than in your dream. In your dream, you commanded, you gave orders, you were obeyed; you were displeased or satisfied, you found your will opposed, you encountered obstacles, you grew angry, you loved, hated, blamed, you came and went. During your medita-tions, hardly were your eyes open in the morning than, possessed anew by the idea that had been occupying you the night before, you would dress, sit at your table, ponder, draw figures, make calculations, eat your dinner, resume your calculations, sometimes getting up from the table to verify them; you would speak to other people, give orders to your servants, eat your supper, go to bed and sleep, without having performed one voluntary action. You have been reduced to a single point; you have acted, but you have not exerted your will. Does one exert will by instinct? Will is always moved by some inward or outward stimulus, by some present impression or recollection of the past, or by some passion or project for the future. After this I need only say one word about freedom, that is, that the most recent action of each one of us is the necessary result of a single cause – oneself; a highly complex cause, but a single one.

Mlle. de l'Espinasse: And necessary?

Bordeu: Undoubtedly. Try to imagine any other action resulting, assuming that the being who acts is the same.

Mlle. de l'Espinasse: He is right. Since I act in a certain way, the person who could act differently is no longer me; and to declare that, at the moment I am doing or saying one thing, I might be saying or doing another, is to declare that I am myself and someone else. But, Doctor, what about vice and virtue? Virtue, so holy a word in all languages, so sacred an idea to all nations!

Bordeu: We must change it for that of doing good, and its contrary for that of doing harm. One is born well or ill endowed by nature; one is irresistibly carried away by the general torrent that brings one man to glory and another to disgrace.

Mlle. de l'Espinasse: What of self-esteem, and shame, and remorse?

Bordeu: Childish reactions founded on the ignorance and vanity of a person who attributes to himself the praise and blame for a moment of time that necessarily had to be.

Mlle. de l'Espinasse: And rewards and punishments?

Bordeu: Ways of correcting that person whom we call wicked, but who can be altered, and of encouraging the one we call good.

Mlle. de l'Espinasse: Isn't there something dangerous about this doctrine?

Bordeu: Is it true or is it false?

Mlle. de l'Espinasse: I believe it to be true.

Bordeu: That is to say, you think that falsehood has its advantages and truth its inconvenient aspects.

Mlle. de l'Espinasse: I think so.

Bordeu: And so do I; but the advantages of falsehood are transient and those of truth are eternal; the distressing results of truth, when they occur, disappear quickly, and those of a lie last as long as the lie. Examine the effects of falsehood in man's mind and in his conduct; in his mind, either falsehood has become somehow or other mingled with truth, and then he is muddle-headed; or else it is thoroughly and consistently united with falsehood, and then he is wrong-headed. Now, what conduct can you expect from a head that is either inconsistent in its reasoning or consistent in its errors?

Mlle. de l'Espinasse: The latter vice, though less contemptible, is perhaps more dangerous than the former.

d'Alembert: Very good; now all is reduced to a question of the faculty of sensation or feeling, memory, organic movements; that suits me very well. But what about imagination? And abstract ideas?

Bordeu: Imagination...

Mlle. de l'Espinasse: One moment, Doctor; let us recapitulate. According to your principles, it seems to me that by a series of purely mechanical operations, I could reduce the greatest genius on earth to an unorganized mass of flesh, which would only retain the faculty of momentary sensation, and that this formless mass could then be brought back from the state of the most utter stupidity imaginable, to the condition of a man of genius. One of these two processes would consist in depriving the original skein of a certain number of its fibres, and thoroughly confusing the rest; and the inverse process, in restoring to the skein the fibres one had removed, and then leaving the whole to a lucky development. For instance: I take away from Newton the two auditory fibres, and he has no more sense of sound; the olfactory fibres, and he has no more sense of smell; the optic fibres, and no more sense of colour; the fibres that form the palate, and he loses his sense of taste; I suppress or entangle the others, and there's an end to the organization of the brain, memory, judgment, desire, aversion, passion, will, self-consciousness, and behold an amorphous mass which has retained only life and sensitiveness.

Bordeu: Two qualities which are almost identical; life pertains to the aggregate, sensitiveness to the elements.

Mlle. de l'Espinasse: I take up this mass again and I restore to it the olfactory fibres, and it can smell; the auditory fibres, and it can hear; the optic fibres, and it can see; the fibres of the palate, and it can taste. Disentangling the rest of the skein, I allow the other fibres to develop, and I behold the rebirth of memory, of the faculty of comparison, of judgment, reason, desire, aversion, passion, natural aptitude, talent, and I find my man of genius once more, without the intervention of any heterogeneous or unintelligible agent.

Bordeu: Excellent; keep to that, all the rest is senseless verbiage. But what about abstract ideas, and imagination? Imagination is the recollection of forms and colours. The picture of a scene or an object inevitably tunes up the sensitive instrument in a certain fashion: either it tunes itself, or it is tuned up by some outside cause. Then it vibrates within, or resounds externally; it retraces in silence the impressions it has received, or echoes them abroad in sounds fixed by convention.

d'Alembert: But its recital exaggerates, omits certain circumstances and adds others, distorts the fact or embellishes it, and the sensitive instruments around it receive impressions which assuredly correspond to those of the instrument which is sounding, but not to the original thing that took place.

Bordeu: True, the recital may be either historical or poetical.

d'Alembert: But how does this poetry or falsehood find its way into the recital?

Bordeu: Because ideas awaken one another, and they awaken one another because they have always been connected. Since you took the liberty of comparing an animal to a harpsichord, you will surely allow me to compare the poet's recital to a song.

d'Alembert: That is quite fair.

Bordeu: In any song there is a scale. This scale has its intervals; each of its notes has its harmonics, and these in turn have their own harmonics. That is how modulations are introduced into the melody, and how the song is enriched and extended. The fact is a given theme that each musician feels in his own way.

Mlle. de l'Espinasse: But why confuse the question with this figurative style? I should say that, since every one has his own eyes, everyone sees and tells a thing differently. I should say that each idea awakens others and that, according to one's turn of mind and one's character, either one keeps to those ideas that strictly represent the fact, or one introduces ideas suggested by association; I should say that there is a choice to be made among these ideas; I should say that this one subject, treated thoroughly, would furnish a whole book.

d'Alembert: You are right; but that won't prevent me from asking the doctor if he is convinced that a form that was not like anything else could not be engendered in the imagination and introduced into the recital.

Bordeu: I think that is the case. The wildest fantasy of this faculty is nothing more than the talent of those tricksters who, from the parts of several animals, compose a strange creature that was never seen in nature.

d'Alembert: And abstract ideas?

Bordeu: They don't exist; there are only habitual omissions, ellipses, that make propositions more general and speech swifter and more convenient. It is the symbols of speech that have given rise to the abstract sciences. A quality common to several beings engendered the terms ugliness and beauty. We first said one man, one horse, two animals; then we said one, two, three, and the whole science of numbers was born.[19] It is impossible to conceive of an abstract word. It was observed that all bodies have three dimensions, length, breadth and depth; each of these was studied, and hence arose all mathematical sciences. An abstraction is merely a symbol emptied of its idea. The idea has been excluded by separating the symbol from the physical object, and it is only when the symbol is attached once more to the physical object that science becomes a science of ideas again; hence the need, so frequently felt both in conversation and in books, of having recourse to examples. When, after a long series of symbols, you ask for an example, you are only requiring the speaker to give body, shape, reality, to attach an idea to the series of sounds made by his speech, by connecting those sounds with sensations that have been experienced.

d'Alembert: Is this quite clear to you, Mademoiselle?

Mlle. de l'Espinasse: Not exceedingly, but the doctor will explain.

Bordeu: You are good enough to say so! No doubt there is some correction and much addition to be made to what I've said; but it is half-past eleven, and at twelve I have a consultation at the Marais.

d'Alembert: Speech swifter and more convenient! Doctor, does one ever understand? Is one ever understood?

Bordeu: Almost all conversations are like accounts already made up... where has my stick got to? One has no idea present in one's mind ... and my hat? And for the simple reason that no man is exactly like another, we never understand precisely, we are never precisely understood; it is always a case of more or less, in everything; our speech always falls short of experience or goes beyond it. A great difference between man's judgments can be observed, an infinitely greater difference passes unobserved, and luckily can never be observed. ... good-bye, good-bye!

Mlle. de l'Espinasse: One word more, I implore you!

Bordeu: Quickly then.

Mlle. de l'Espinasse: Do you remember those leaps of which you spoke to me?

Bordeu: Yes.

Mlle. de l'Espinasse: Do you think that fools and men of intelligence might have those leaps in their lineage?

Bordeu: Why not?

Mlle. de l'Espinasse: All the better for our great-nephews; perhaps a second Henri IV will appear.

Bordeu: Perhaps he has already appeared.

Mlle. de l'Espinasse: Doctor, you must come and dine with us.

Bordeu: I'll do what I can, I don't promise: expect me when you see me.

Mlle. de l'Espinasse: We will wait for you till two o'clock.

Bordeu: So be it.

CONCLUSION OF THE CONVERSATION

The Speakers: *Mademoiselle de l'Espinasse and Doctor Bordeu.*

About two o'clock the doctor came back. D'Alembert had gone out to dine, and the doctor was alone with Mlle. de l'Espinasse. Dinner was served. They talked of indifferent matters until dessert; but when the servants had retired, Mlle. de l'Espinasse said to the doctor: Come now, Doctor, drink a glass of malaga, and then you shall give me the answer to a question that has passed through my head a hundred times, and that I shouldn't dare put to anyone but you.

Bordeu: Excellent malaga, this. What is your question?

Mlle. de l'Espinasse: What do you think of the intermingling of species?

Bordeu: Well, that is certainly a good question. I think that men have attributed great importance to the act of generation, and rightly so; but I'm not satisfied with their laws, either civil or religious.

Mlle. de l'Espinasse: And what fault do you find with them?

Bordeu: They have been made without justice, without purpose, and without any consideration for the nature of things or for the public good.

Mlle. de l'Espinasse: Try and explain.

Bordeu: I mean to... But wait... (*he looks at his watch*) I've still a good hour to spare you; I'll go quickly, and it will be long enough. We are alone, you're no prude, you won't fancy that I intend any lack of that respect I owe you; and whatever may be your opinion of my ideas, I hope, on my side, that you won't conclude therefrom anything derogatory to my morals.

Mlle. de l'Espinasse: Of course not; but I don't much like your opening.

Bordeu: In that case let's change the subject.

Mlle. de l'Espinasse: No, no, go on. One of your friends who was looking out for husbands for myself and my two sisters allotted a sylph to the younger, a great angel of the Annunciation to the elder and a disciple of Diogenes to me; he knew us well, all three. Nevertheless, Doctor, a veil, just a slight veil.

Bordeu: That goes without saying, insofar as the subject and my profession allow of it.

Mlle. de l'Espinasse: It won't inconvenience you. But here is your coffee ... drink your coffee.

Bordeu: (having drunk his coffee): Your question has physical, moral and poetical aspects.

Mlle. de l'Espinasse: Poetical!

Bordeu: Surely; the art of creating non-existent beings in imitation of those that exist is true poetry. This time then, instead of Hippocrates, you'll allow me to quote Horace. This poet, or maker, says somewhere: *Omne tulit punctum, qui miscuit utile dolci*; the supreme merit lies in combining the pleasant with the useful. Perfection consists in reconciling these two qualities. The action that is both pleasant and useful must occupy the first place in the aesthetic hierarchy; we cannot deny the second place to that which is useful; the third will be for what is pleasant; and to the lowest rank we must relegate the action that produces neither pleasure nor profit.

Mlle. de l'Espinasse: So far I can agree with you without blushing. Bur where is it going to lead us?

Bordeu: You shall see. Mademoiselle, can you tell me what profit or what pleasure is derived from strict chastity and continence, either by the individual who practises them or by society?

Mlle. de l'Espinasse: None, I declare.

Bordeu: Then, despite the magnificent praises lavished on them by fanaticism, despite the protection afforded them by civil laws, we will cross them out of the catalogue of virtues, and we will agree that there is nothing so childish, so ridiculous, so absurd, so harmful, so contemptible, nothing worse, except positive evil, than these two rare qualities.

Mlle. de l'Espinasse: I'll grant you that.

Bordeu: Take care, I warn you, you'll want to withdraw in a moment.

Mlle. de l'Espinasse: We never withdraw.

Bordeu: And what about solitary actions?

Mlle. de l'Espinasse: Well?

Bordeu: Well, they at least give pleasure to the individual, and either our principle is wrong or. ...

Mlle. de l'Espinasse: What, Doctor! ...

Bordeu: Yes, Mademoiselle, yes, since on the one hand they are just as neutral, and on the other they are not so sterile. It is a need, and even if one were not urged by the need it is still a pleasant experience. I want people to be well, I absolutely insist on it, do you understand? I am against all excess, but, in a state of society such as ours, there are a hundred reasonable considerations if there's one, such as lack of fortune, the dread, for men, of a painful repentance, for women the dread of dishonour; not to mention passionate temperament and the disastrous effects of strict continency, particularly on young people, which all drive a wretched creature that's consumed with languor and boredom, a poor devil who doesn't know where to get help, to relieve himself in the manner of the cynic. Would Cato, who said to a young man on the point of visiting a courtesan: 'Courage, my son...' speak to him in the same way to-day? If, on the contrary, he caught him in the act alone, would he not add: 'That is better than corrupting the wife of another, or risking one's honour and one's health'? What! Just because circumstances deprive me of the greatest happiness imaginable, that of mingling my senses with those of a partner chosen by my heart, my ecstasy with her ecstasy, my soul with her soul, and of reproducing myself in her and with her; just because I cannot imprint upon my action the sacred stamp of utility, must I forbid myself the enjoyment of a necessary and delicious moment? One is bled to relieve plethora; and what matters the nature of the superabundant humour, its colour, and the way one gets rid of it? It is just as superfluous in the one disturbance as in the other; and if it were pumped back out of the vessels that contain it, distributed throughout the whole body, to find its way out by a longer, more painful and perilous way, would it be any the less wasted? Nature allows nothing useless; and how can I be held guilty for helping her, when she appeals for my assistance by the plainest of symptoms? Let us never provoke her, but let us lend her a hand when the occasion demands it; to refuse this, to remain idle, seems to me mere foolishness and a lost chance of pleasure. Live soberly, people may say to me, tire yourself out. I understand: I am to deprive myself of one pleasure: I am to inflict pain on myself to ward off another pleasure. A very happy notion!

Mlle. de l'Espinasse: Your doctrine isn't suitable to be preached to children.

Bordeu: Nor to anyone else. Nevertheless, will you allow me to suggest a possibility? You have a daughter who is virtuous, too virtuous; innocent, too innocent; she has reached the age when the temperament develops. Her mind becomes bewildered, nature does not assist her; you send for me. I see at once that all the symptoms that alarm you arise from the superabundance and retention of the seminal fluid; I warn you that she is threatened with a kind of madness that is difficult to prevent and sometimes impossible to cure; I indicate the remedy. What will you do?

Mlle. de l'Espinasse: To tell you the truth, I believe... but such cases don't occur.

Bordeu: That's where you are wrong; they are not uncommon, and they would be quite common if the looseness of our morals did not prevent it. Be that as it may, to divulge such principles would mean trampling underfoot all decency, exposing oneself to the most odious suspicions and offending the dignity of society. But you're absorbed by some thought.

Mlle. de l'Espinasse: Yes, I was wondering if I should ask you whether you had ever had to impart this secret to any mothers?

Bordeu: Certainly.

Mlle. de l'Espinasse: And what course did these mothers take?

Bordeu: All, without exception, took the right course, the sensible course. I would not take off my hat in the street to a man suspected of practising my doctrine; it would be enough for me that he'd be called a vile wretch. But we are talking without witnesses and informally; and I will say to you about my philosophy what Diogenes, stark naked, said to the young and bashful Athenian with whom he was preparing to wrestle: 'My son, fear nothing, I am not so wicked as yonder man.'

Mlle. de l'Espinasse: Doctor, I see where you are tending, and I wager...

Bordeu: I won't wager, you would win. Yes, Mademoiselle, such is my opinion.

Mlle. de l'Espinasse: What! Whether one remains within the limits of one's own species, or passes beyond them?

Bordeu: True.

Mlle. de l'Espinasse: You are monstrous!

Bordeu: Not I, but either nature or society. Listen, Mademoiselle, I don't let myself be imposed on by words, and I express myself all the more freely because my conscience is clear, and the purity of my morals beyond reproach on all sides. I will therefore ask you: of two actions both confined solely to pleasure, which can only bring enjoyment without profit, but of which one brings enjoyment to the agent alone, whereas in the other the enjoyment is shared by the agent and a fellow-creature, male or female, for the sex and even the use of sex do not affect the question, in favour of which will common sense declare itself?

Mlle. de l'Espinasse: Such questions are too lofty for me.

Bordeu: Oh! After having been a man for four minutes, now you're resuming mob-cap and petticoats, and becoming a woman again. Very good; well! you shall be treated as such. That is easily done. We hear nothing nowadays about Madame du Barry. You see, everything is settling down; people thought the court would be turned upside down. The master acted like a sensible man; *'omne tulit punctum';* he's kept the woman who gave him pleasure, and the minister who was useful to him. But you are not listening to me. Where have you got to?

Mlle. de l'Espinasse: I'm thinking of those unions that seem to me wholly against nature.

Bordeu: Nothing that exists can be either against nature or outside nature. I don't except even voluntary chastity and continence, which would be the chief crimes against nature if one could sin against nature, and the chief crimes against the social laws in a country where actions were weighed in a balance other than that of fanaticism and prejudice.

Mlle. de l'Espinasse: I am back at your accursed syllogisms and I see no middle course, one has to deny or accept everything. But see, Doctor, the most honest way and the quickest, is to jump over the mess and come back to my first question: What do you think of the intermingling of species?

Bordeu: We don't need to jump to get there; we were there already. Do you mean from the physical or the moral point of view?

Mlle. de l'Espinasse: Physical, physical.

Bordeu: So much the better: the moral question came first and you've decided it. So then...

Mlle. de l'Espinasse: Agreed... no doubt it is a preliminary, but I wish that you could separate cause from effect. Let us leave the horrid cause out of it.

Bordeu: You are asking me to begin at the end; but, since you desire it, I will tell you that, thanks to our faintheartedness, our aversions, our laws, our prejudices, very few experiments have been made. It is not known in which cases copulation would be wholly unfruitful, in which cases utility and pleasure would combine; what sort of species might be expected from varied and continuous experimentation; whether fauns are real or fabulous creatures; whether we could not multiply races of mules in a hundred different ways, and whether those that we know are really sterile. But there's one odd fact which an infinite number of learned folk will swear to you is true, and which is false; that is, that they have seen in the archduke's poultry-yard a vile rabbit acting as cock to twenty vile hens, who put up with him. They will add that they were shown chickens covered with hair, the product of this bestiality. You may take it from me that someone was making fun of them.

Mlle. de l'Espinasse: But what do you mean by continuous experimentation?

Bordeu: I mean that the circulation of creatures is gradual, that their assimilation has to be prepared beforehand, and that, in order to succeed in such experiments, one ought to start a long way back and endeavour first to make animals more like one another by a similar diet.

Mlle. de l'Espinasse: You'll find it hard to bring a man to graze.

Bordeu: But not to drink goat's milk frequently, and one could easily bring the goat to feed on bread. I've chosen the goat for reasons peculiarly my own.

Mlle. de l'Espinasse: What are these reasons?

Bordeu: You are very bold! They are… well… they are, that we should thus produce a vigorous, swift, intelligent and indefatigable race of beings, of whom we could make excellent servants.

Mlle. de l'Espinasse: Very fine, Doctor; I fancy already that I can see five or six great insolent goats'-feet behind the carriages of your duchesses, and it delights me.

Bordeu: And we should no longer degrade our brothers by subjecting them to functions unworthy of them and of ourselves.

Mlle. de l'Espinasse: Better still.

Bordeu: And that we should no longer reduce men in our colonies to be mere beasts of burden.

Mlle. de l'Espinasse: Quickly, quickly, Doctor, set to work and make these goats'-feet for us.

Bordeu: You have no scruples about allowing it?

Mlle. de l'Espinasse: Stop, though, one has occurred to me; your goats'-feet would be wildly licentious.

Bordeu: I can't guarantee they'd be highly moral.

Mlle. de l'Espinasse: There will be no more safety for honest women; they will multiply unceasingly, and in the end we shall have either to destroy them or obey them. I don't want them any more, I don't want them any more. You had better keep quiet.

Bordeu: (going away): And the question of their baptism?

Mlle. de l'Espinasse: Would cause a great to-do in the Sorbonne.

Bordeu: Have you seen in the King's garden, in a glass cage, an orangoutang that looks like St John preaching in the wilderness?

Mlle. de l'Espinasse: Yes, I've seen it.

Bordeu: Cardinal de Polignac said to it one day: 'Speak, and I will baptize thee.'

Mlle. de l'Espinasse: Good-bye then, Doctor; don't forget us for centuries as you do, and remember sometimes that I love you to distraction. If people only knew what horrors you've been telling me!

Bordeu: I'm sure you'll keep silent about them.

Mlle. de l'Espinasse: Don't be too confident, I only listen for the pleasure of repeating things. But just one more word, and I'll never reopen the subject again.

Bordeu: What is it?

Mlle. de l'Espinasse: Whence come these abominable tastes?

Bordeu: Everywhere from a weakness of the organism among young people and the mental corruption of the old; in Athens, from the attraction of beauty; in Rome, from the scarcity of women; and in Paris, from the fear of the pox. Good-bye, good-bye.

(1769)

Notes

D'Alembert's Dream offers particular difficulties in translation because Diderot had to employ a certain limited number of words to describe several different genetic, embryological and histological phenomena, of which a clear conception did not then exist in science, and for which the requisite precise technical terms were therefore not available. Diderot was reaching forward and suggesting hypotheses in connection with these ideas, which were then no doubt the subjects of discussion among the scientists among whom Diderot moved. In the light of later scientific development it is possible to understand what were the conceptions which Diderot was seeking to express. In the text the terminology has been left as Diderot wrote it; and the notes can indicate the different senses in which Diderot used the same few nontechnical words.

1 'I believe I have told you that I had made a dialogue between d'Alembert and myself. On re-reading it, it took my fancy to make a second one, and it has been done. The characters in it are d'Alembert who dreams, Bordeu and the friend of d'Alembert, Mlle. de l'Espinasse. It is called *D'Alembert's Dream*. It isn't possible to be more profound and more extravagant. I added afterwards five or six pages that would make my lover's hair stand on end; but, however, she will never see them! But this will surprise you very much, that there's not a word about religion and nor a single improper word. After that I defy you to guess what it can be.' (*Œvres Complètes*, Vol. XIX, p. 315. *Lettres à Mlle. Volland*, No. 124, 2 September, 1769.)

The 'five or six pages' that would make Sophie Volland's 'hair stand on end,' refers to the *Continuation of the Conversation* (pp. 133–9).

2 Diderot says: '*Une molécule sensible et vivante se fond dans une molécule sensible et vivante.*' It is best to translate *molécule* here as particle.

3 '*Life is the mode of existence of albuminous substances*, and this mode of existence essentially consists in the constant self-renewal of the chemical constituents of these substances. The term albuminous substance is used in the sense used by modern chemistry, which includes under this name all substances constituted similarly to ordinary white of egg, otherwise also known as protein substances. The name is inappropriate, because ordinary white of egg plays the most lifeless and passive role of all the substances related to it, since, together with the yolk, it is merely food for the developing embryo. But while so little is as yet known of the chemical composition of albuminous substances, this name is yet better than any because it is more general.'

'Everywhere where we find life we find it associated with an albuminous body, and everywhere where we find an albuminous body not in process of dissolution, there also, without exception, we find the phenomena of life...'

'But what are these universal phenomena of life which are equally present among all living organisms? Above all, an albuminous body absorbs other appropriate substances from its environment and assimilates them, while other, older parts of the body are consumed and excreted. Other, non-living, bodies also change, and are consumed or enter combinations in the course of natural processes; but in doing this the cease to be what they were. A rock worn away by atmospheric action is no longer a rock; metal which oxidizes rusts away. But what with non-living bodies is the cause of destruction, with albumen is *the fundamental condition of existence*. From the moment when the uninterrupted metamorphosis of its constituents, this constant alternation of nutrition and excretion [cf. Diderot's 'continual action and reaction' – *Ed.*] no longer takes place in an albuminous body, from that moment the albuminous body itself comes to an end and decomposes, that is, dies. Life, the mode of existence of albuminous substance, therefore consists primarily in the fact that at each moment it is itself and at the same time something else; and this does not take place as the result of a process to which it is subjected from without, as is the way in which this can occur in the case of inanimate bodies. On the contrary, life, the exchange of matter which takes place through nutrition and excretion, is a self-completing process which is inherent in and native to its medium, albumen, without which it cannot exist' (F. Engels, *Anti-Dühring*, pp. 94–6).

(By albuminous substance Engels does not mean a protein in its modern sense as a pure crystalline chemical substance, but the complex chemicals that underlie protoplasm: proteins, carbohydrates, lipides, salts. – *Ed.*)

4 The 'Eel-man'; name given by Voltaire to Needham (1713–1781), an English biologist and Catholic divine, who believed in the spontaneous generation of little 'eels' from fermenting flour. He really saw the micro-organisms which were causing the fermentation. In 1745 Needham published *'An Amount of some Miroscopical Discoveries founded on an Examination of the Clamary and its Wonderful Milt vessels.'*

5 Diderot uses the word *'atome'* obviously not in the chemical sense.

6 Cf. Engels's discussion of time and space in *Anti-Dühring*. He first quotes Kant's antinomy concerning time and space. *The First Antinomy of Pure Reasons* has as thesis: The world has a beginning in time and limited also with regard to space. Kant then gives the proof of it. He then states and proves the contrary thesis: the world can have no beginning in time and no end in space. (Kant, *Critique of Pure Reason*, Part I, Section II, Book II, Div II, §2. English tr. Max Müller, p. 344, 346). In this he finds the antimony, the insoluble contradiction, that the one thesis is just as demonstrable as the other. Engels continues: 'The problem itself has a very simple solution. Eternity in time, infinity in space, mean from the start, and in the simple meaning of the words, that there is no end in any direction, neither forwards nor backwards, upwards or downwards, to the right or to the left. This infinity is something quite different from that of an infinite series, for the latter always starts out from one, with one first term...' (*Anti-Dührung*, pp. 60–1).

Engels then discusses 'infinite series' in space and time: 'It is clear that the infinity which has an end but no beginning is neither more nor less infinite than that which has a beginning but no end. The slightest dialectical insight ... [would show] ... that beginning and end are necessarily interconnected like the North Pole and the South Pole, and that if the end is left out, the beginning just becomes the end – the one end which the series has; and *visa versa*. The whole fraud would be impossible but for the mathematical usage of working with infinite series. Because in mathematics it is necessary to start from definite, finite terms in order to reach the indefinite, the infinite, all mathematical series, positive or negative must start from 1, or they cannot be used for calculation. The abstract requirements of a mathematician are, however, very far from being a compulsory law of the world of reality.

'Infinity is a contradiction, and is full of contradictions. From the outset it is a contradiction that an infinite is composed of nothing but finites, and yet this is the case. The finiteness of the material world leads no less to contradictions than its infiniteness, and every attempt to get over these contradictions leads ... to new and worse contradictions. It is just *because* infinity is a contradiction that it is an infinite process, unrolling endlessly in time and space. The removal of the contradiction would be the end of infinite' (*Anti-Dühring*, p. 62).

'From the dialectical standpoint, the possibility of expressing motion in its opposite, in rest, presents absolutely no difficulty. To dialectical philosophy the whole contradiction, as we have seen, is only relative; there is no such thing as absolute rest, unconditional equilibrium, and the motion as a whole puts an end to the equilibrium. When therefore rest and equilibrium occur they are the result of arrested motion, and it is self-evident that this motion is measurable in its result, and can be expressed in it, and can be resorted out of it again in one form or another' (*Anti-Dühring*, p. 74).

7 Epicurus (341–271 BC) was one of the great Greek materialists. He developed and extended the theories of Democritus. The work of Epicurus dominates the history of early materialism. Besides relatively fragmentary remains of Epicurus's writings, the best exposition of ideas is given in the poem *De Rerum Natura* (Concerning the Nature of Things) by the Latin poet Lucretius (*c.* 97–55 BC).

Epicurus founded a science of the universe with a materialist basis, eliminating the powers and activities of spirits, without supernatural laws and without any 'heavenly justice' towards man.

The theory of knowledge of Epicurus established the primacy of the external world in opposition to philosophic idealism, particularly that of Plato. Knowledge, perception, is an 'action' of the external world on man, although the mind is not entirely passive before it.

On this basis Epicurus constructed his physics. He took over from Democritus the atomic conception of all that exists; the motion of the hard, indivisible atoms, as they move to occupy the non-resisting void between them, creates the universe; a conception of the world where everything happens by virtue of mechanical casuality, chance and finality being excluded. Epicurus developed this doctrine in relation to the properties and movements of the atoms. He attributed weight to the atoms, limited and not unlimited shapes, and a contingent source of motion, 'declination', in addition to motion as a result of juxtaposition to unresisting void. Epicurus envisaged a universe without finality, without providence or destiny, where only mechnical causes operated and where the soul even, and the gods, were described as complex structures of material atoms.

8 No case of supposed spontaneous generation of living organisms from non-living matter has yet been proved; alleged cases have been shown to be due to fortuitous infection by micro-organisms. This was demonstrated in particular by Pasteur, in experiments designed to show that alleged cases of spontaneous generation did not take place under sterile conditions. Nevertheless, modern science believes in the chemical origin of life at a definife period in the Earth's development.

'Through combination of modern biochemical knowledge with astrophysical and geological considerations about the early atmosphere of the planet, we can make a plausible picture of the origin of life by purely chemical means, and no other hypothesis for its origin can be put forward which will stand the slightest rational examination' (O. D. Bernal, *Engels and Science*).

9 Father Castel's ribbon. This was an instrument, an 'ocular clavecin' invented by Father Castel in which multicoloured ribbons were combined in colour harmonies by striking a keyboard.

10 Compare with Lenin's *Elements of Dialectics*:

'2. *Totality* of the manifold *relations* of the things to others.

'8. The *relations* (of the thing or appearance) not only manifold but *general, universal.*

Everything (appearance, process, etc.) is connected with *every other*.' (Quoted in T. A. Jackson, *Dialectics*, p. 635.)

11 Cf. Note 3.

12 'The gross developments of a network (*réseau*) that forms itself, increases, extends, and throws out a multitude of imperceptible threads (*fils*),' i.e., the development of the nervous system.

13 The subsequent passage where Bordeu is speaking is not easy to understand. It is evident, however, that Diderot is attempting to describe the development of the fertilized egg cell and subsequent cell-differentiation. With this is intermingled a conception of the development of the nervous system. To express these conceptions only a few non-technical words were used by Diderot:

(a) '... that speck became a loose thread, then a bundle of threads,' (*ce point devint un fil délié, puis un faisceau de fils*). The 'bundle of threads' here suggests the collection of cells at early stages of growth of the egg, rather than an allusion to nerve fibres, which is the sense in which it is used again later.

(b) 'Each of the fibres of the bundle of threads (*chacun des brins du faisceau de fils*) was transformed, solely by nutrition and according to its conformation (*par la seule nutrition et par sa conformation*) into a particular organ; exception being made of those organs in which the fibres of the bundle are metamorphosed, and to which they give birth,' (*abstraction faite des organes dans lesquels les brins du faisceau se métamorphosent, et auxquelles ils donnent naissance*). This appears to refer to the organs, the gonads, where the germ-cells themselves are reproduced. The word '*brin*' meaning literally shoot, sprig, blade (of grass), stick, bit, is clearly used to mean a sub-division of thread (*fil*). The use of '*brin*' also meaning the *staple* (of rope, etc.) gives a clue justifying translation as *fibre*, which we have used, as being the unit from which a thread is made, by taking many fibres together.

(c) 'The bundle is a purely sensitive system.' The bundle (*faisceau*) here seems to refer to the nervous system.

(d) 'Fibre' (*brin*) here refers to the various nerves.

(e) '... bundle with a peculiar fibre which would give rise to an organ unknown to us.' (*un faiseau avec un brin singulier qui donnerait naissance à une organe qui nous est intonnu.*)

(f) '...two fibres which characterize the two sexes.' (*... les deux brins qui charactérisent les deux sexes.*)

If the 'fibres' (*brins*) in phrases *e* and *f* are thought of in this context as the chromosomes of the cells, which control the subsequent development and sex of the organism, the whole passage may be looked upon as another example of the way in which Diderot was seeking to develop a purely materialistic hypothesis, entirely on theoretical grounds in advance of scientific knowledge, to account for the 'mysteries' of biological development. All this has a surprisingly modern ring. (g) Here 'fibre' (*brin*) refers once more to the nerve fibres.

14 Diderot is suggesting how interference with the germ-cells or embryo might alter the structure of the adult. If the subsequent passage, dealing with abnormalities, etc., is read also bearing in mind the chromosome conception (the chromosomes being the cell structures which carry the genes controlling the development of the organism, the genetic material) then this passage is seen to be a brilliant fore-shadowing, purely theoretical at that time, of subsequent ideas of biological development and heredity.

15 The remarks in the previous note, about the germ-cells and chromosomal control of the adult structure apply here also.

The French is: '*Pour fair un enfant on est deux, comme vous savez. Peut-être qu'un des agents répare le vice de l'autre, et que le réseau défectueux ne renaît que dans le moment où le descendant de la race monstrueuse prédomine et donne le loi à la formation du réseau. Le faisceau de fils constitue la différence originelle et première de toutes les espèces d'animaux. Les variétés du faisceau d'une espèce font toutes les variétés monstrueuses de cette espèce.*'

In this passage, if the 'bundle of threads' (*faisceau de fils*) is understood as the chromosomes of the germ-cells, Diderot's genetical hypothesis is quite straightforward. The 'leaps' could be mutations.

16 '... balloons under his feet.' In *tabes dorsalis*, the soles of the feet lack sensation, and hence there is the feeling of an inert layer between where feeling ends in the foot and where the sole actually touches the ground.

17 *Sensibilité* seems best translated by sensibility or feeling here, since it is referring to a general, temperamental or intellectual condition, and not to the specific sensitiveness of living matter.

18 This is a brilliant 'interpretation of dreams.'

19 The concepts of number and form have not been derived from any source other than the world of reality. The ten fingers on which men learnt to count, that is, to carry out the first arithmetical operation, may be anything else, but they are certainly not a free creation of the mind. Counting requires not only objects that can be counted, but also the ability to exclude all properties of the objects considered other than their number – and this ability is the product of a long historical evolution based on experience. Like the idea of number, so the idea of form is derived exclusively from the external world, and does not arise in the mind as a product of pure thought. There must be things which have shape and whose shapes are compared before anyone can arrive at the idea of form....

'Before it was possible to arrive at the idea of deducing the *form* of a cylinder from the rotation of a rectangle about one of its sides a number of real rectangles and cylinders, in however imperfect a form, must have been examined. Like all other sciences, mathematics arose out of the *needs* of men; from the measurement of land and of the content of vessels, from the computation of time and mechanics....

'The ideas of lines, planes, angles, polygons, cubes, spheres, etc., are all taken from reality, and it requires a pretty good portion of naïve ideology to believe the mathematicians – that the first line came into existence through the movement of a point in space, the first plane through the movement of a line, the first solid through the movement of a plane and so on. Even language rebels against such a conception. A mathematical figure of three dimensions is called a solid body, *corpus solidum*, hence even in Latin a tangible object; it has therefore a name derived from sturdy reality and by no means from the free imagination of the mind.' (F. Engels. *Anti-Dühring*, pp. 47–9.)

Letter on the Blind for the Use of those who See

Letter on the Blind for the Use of those who See[1]

Possunt nec posse videntur.
(Virgil, Aeneid, liber v, 23)[2]

It was not more than I suspected, that the blind girl whom Monsieur de Réaumur had couched for cataract would not inform you of what you were anxious to know; but I little thought it would be neither her fault nor yours. I have in person, and by means of his best friends and by paying him many compliments, applied to her benefactor, but all in vain; the first dressing will be removed without you. Some persons of the highest distinction have had the honour of sharing this refusal with philosophers, and, in a word, he does not wish to remove the veil, except in the presence of some eyewitnesses of no great importance. If you would know why that wonderful operator makes a secret of experiments at which you think too great a number of intelligent witnesses cannot be present, my answer is, that the observations of such a celebrated person do not so much stand in need of spectators, whilst making, as of hearers when made. Thus, disappointed, madam, I have returned to my original intention, and, since I was forced to go without an experiment in which I saw little profit would accrue to you or to me, but of which Monsieur de Réaumur will doubtless make a much better use, I set to work to philosophise with my friends upon the important matter which is the object of it. How happy should I be, if the narrative of one of our conversations might stand instead of the spectacle I so rashly promised you! The day that the Prussian[3] operated on Simoneau's daughter for a cataract, we went talk with the Puisaux[4] man who was born blind. He is possessed of good solid sense, is known to great numbers of persons, understands a little chemistry, and has attended the botanical lectures at the Jardin du Roi with some profit to himself. His father was a distinguished professor of philosophy at the University of Paris. He had private means, sufficient to have satisfied his remaining senses, but a taste for pleasure led him into some excesses in his youth; people took advantage of his weaknesses, his affairs became embarrassed, and finally he withdrew to a little town in the provinces, from which he pays a yearly visit to Paris, bringing with him liqueurs which give great satisfaction. These, madam, are not very philosophic details, but for that very reason are likely to convince you that the person I am speaking of is not imaginary.

We arrived at our blind man's house about five o'clock in the afternoon, and we found him busy teaching his son to read with raised letters. He had only been up an hour, for I must tell you the day begins for him when it is ending for us. His custom is to look after his household affairs and to work while others are asleep. At midnight nothing interrupts him, and he is in no-one's way. His first care is to set in its place everything that has been displaced during the day, and when his wife wakes she generally finds the house tidy. The difficulty the blind have in finding things that are mislaid makes them orderly, and I have observed that their intimates also share this quality, either from the effect of the good example of the blind, or from a feeling of compassion towards them. How unhappy would the blind be without the little attentions of those about them! – nay, we ourselves feel the lack of them. Great services are like the large gold or silver coins that we rarely make use of, but small attentions are small change which is always passing from hand to hand.

This blind man is a good judge of symmetry. Symmetry, which is perhaps a matter of pure convention among us, is certainly so in many respects between a blind man and the sighted. A blind man studies by his touch that disposition required between the parts of a whole to enable it to be called beautiful; and then at length attains to a just application of that term. But in saying 'that is beautiful,' he does not form an opinion, it is no more than repeating the judgment of those who see; and is not this the case of three quarters of those who give their opinion on a play or a book? Beauty for the blind is but a word when divorced from utility, and, lacking an organ, how many things are there the utility of which escapes them? Are the blind not very much to he pitied in accounting nothing beautiful unless it be likewise good? How many admirable things are lost to them! The only compensation for their loss is that their ideas of beauty, though less extensive, are more definite than those of many keen-sighted philosophers who have written prolix treatises on the subject. This blind man often speaks of mirrors. You think he does not know the meaning of the word, yet he is never known to put a glass in a wrong light. He speaks as sensibly as we do on the qualities and defects of the organ which he lacks. Even if he attaches no idea to the terms he makes use of, still he has the advantage over most other men that he never uses them wrongly. He speaks so wisely and so well of so many things absolutely unknown to him, that his conversation would considerably lessen the weight of that inference which, without knowing the reasons, we all draw from what passes in ourselves to what passes within the minds of others.

I asked him what he meant by *a mirror*? 'An instrument,' answered he, 'which sets things in relief at a distance from themselves, when properly placed with regard to it. It is like my hand, which, to feel an object, I must not reach to one side of it.'

Had Descartes been born blind, he might, I think, have hugged himself for such a definition. Pray consider what an ingenious combination of ideas it implies. This blind man's only knowledge of objects is by touch. He knows by hearing other men say so that they know objects by sight as he knows them by touch; at any rate that is the only idea he can form of the process. He also knows that we cannot see our own face though we can touch it. Sight, he therefore concludes, is a kind of touch which extends to distant objects and is not applied to our face. Touch gives him an idea only of relief. Therefore, he concludes, a mirror is an instrument that represents us in relief outside ourselves. How many famous philosophers have laboured with less subtlety to arrive at conclusions equally erroneous! But if a mirror astonished our blind man, how much greater was his surprise when we told him that there are instruments which magnify objects, while others remove them without duplicating them, put them out of their place, bring them nearer, remove them farther, and reveal the minutes' details to the eyes of naturalists; while others again multiply objects a thousand times, and others appear to change the figure of objects completely. He asked us a hundred curious questions concerning these phenomena. For instance, he asked us if only persons who were called naturalists could see with the microscope, and if only astronomers could see with the telescope; if the instrument for enlarging objects were bigger than that for diminishing them; if that which brings them nearer were shorter than that for diminishing them. But what puzzled him was that the other self, which according to him the mirror represents in relief, should not be tactile.

'So this little instrument,' said he, 'sets two senses to contradict one another; a more perfect instrument would perhaps reconcile these contradictions, without the object being ever more real for that, and perhaps a third instrument, still more perfect and less illusory, would cause these contradictions to disappear and show us our error.'

'And what are eyes, do you suppose?' asked Monsieur d'——. 'An organ,' replied the blind man, 'on which the air has the effect this stick has on my hand.' That answer amazed us, and while we gazed at one another in astonishment he continued: 'When I place my hand between your eyes and an object, my hand is present to you but the object is absent. The same thing happens when I reach for one thing with my stick and come across another.'

Madam, only turn to Descartes' *Dioptrics*, and there you will see the phenomena of sight illustrated by those of touch, and the plates full of men busied in seeing with sticks. Descartes, and all the later writers, have not been able to give us clearer ideas of vision; and that great philosopher was, in this respect, no more superior to the blind man than a common man who has the use of his eyes.

No one thought of asking him questions as to painting and writing, but it is obvious that his comparison would fit in with every question, and I make no doubt but that he would have told us that to try to read or to see without eyes was like looking for a pin with a thick stick. We only talked to him about those kinds of glasses which exhibit objects in relief, and which are both so very similar to and so very different from mirrors; but these we perceived contradicted rather than coincided with his idea of a looking-glass, and he was apt to think that a painter might perhaps paint a looking-glass, and thus it came to represent objects in colours.

We saw him thread very fine needles. May I ask you, madam, to suspend your reading for a while and try what you would do in his place? In case you do not light upon any suitable means, I will tell you of our blind man's. He takes the eye of the needle between his lips in the same direction as his mouth, then by his tongue and suction he draws in the thread, which follows his breath unless it is much too thick for the eye; but in that case a man with sight is in the same difficulty as the blind.

He has a surprising memory for sounds, and can distinguish as many differences in voices as we can in faces. He finds in these an infinite number of delicate gradations which escape us because we have not the same interest in observing them. For us, these shades of difference are like our own countenance. Of all the men we have seen, the one we least remember is our own self. We notice faces to recognise people; and if we do not remember our own, it is because we are never liable to mistake ourselves for another person or another for ourselves. Moreover, the mutual aid our senses lend stands in the way of their perfection. This will not be the only occasion where I shall have to remark upon this.

On this point our blind man said: 'That he should think himself a pitiable object in wanting those advantages which we enjoy, and that he should have been inclined to consider us as superior beings if he had not a hundred times found us very much inferior to him in other respects.' This reflection led to another. This blind man, we said, values himself as much as, and perhaps more than, we who see. Why then, if the *brute* reasons (and it is scarce to be doubted), why if it knows its its advantages over man better than those of man over it, should it not make a similar inference? He has arms, perhaps says the gnat, but I have wings. He has weapons, says the lion, but have I not claws? The elephant would look on us as insects; and all the animals, while allowing us reason, with which we should at the same time stand in great need of their instinct, would claim that with their instinct they could do very well without our reason. We have such a strong desire to exaggerate our qualities, and make little of our defects, that it would seem man's part to write a treatise on force, and animals' on reason.

One of our company thought to ask our blind man if he would like to have eyes. 'If it were not for curiosity,' he replied, 'I would just as soon have long arms: it seems to me my hands would tell me more of what goes on in the moon than your eyes or your telescopes; and besides, eyes cease to see sooner than hands to touch. I would be as well off if I perfected the organ I possess, as if I obtained the organ which I am deprived of.'

Our blind man points with such exactness at the place a noise comes from that I make no doubt the blind may, by practice, become very dexterous and very dangerous. I will tell you a story which will convince you how imprudent it would be to stand the throwing of a stone or firing a pistol by a blind man, if he was in the least used to that weapon. He had in his youth a quarrel with one of his brothers, who came off badly. Provoked at some insulting language, he seized the first missile which came to hand, threw it at him, and hit him directly on the forehead, and laid him flat on the ground.

This, with some other occurrences of a similar kind, caused him to be brought before the police. The outward show of power, which affects us so strongly, is nothing to the blind. Our blind man appeared before the magistrate, as before an equal; menaces did not intimidate him. 'What will you do to me?' he asked Monsieur Herault.[5] 'I will commit you to a dungeon,' answered the magistrate. 'Ah, sir,' the blind man replied, 'I have been in one for twenty-five years.' There was an answer, madam; and what a text for one who is so fond of moralising as your humble servant! We quit life as we would a charming scene, the blind leave it as a dungeon; and if we have more pleasure in living than him, he has less reluctance to meet his end.

The blind man of Puisaux judges his proximity to the fire by the degrees of heat; of the fullness of vessels by the sound made by liquids which he pours into them; of the proximity of bodies by the action of the air on his face. He is so sensitive to the least atmospheric change, that he can distinguish between a street and a closed alley. He is an extremely good judge of the weight of bodies and the capacity of vessels; and he has trained his arms to be such an exact balance, his fingers to be such skilful compasses, that in this kind of statics I would always back our blind man[6] against twenty people with all their eyes about them. The smooth surface of bodies has as many shades of difference for him as the sound of voices, and there is no risk of his mistaking his wife for another, unless he was to profit from the mistake. Yet it is very probable that among a blind people wives would be in common, or their laws against adultery must be severe indeed, so simple would it be for wives to deceive their husbands by concerting a sign with their gallants.

He judges beauty by touch – that is easy to understand; but what is not so easy to grasp is that his judgment is influenced by pronunciation and the sound of a voice. Anatomists ought to tell us if there is any relation between the parts of the mouth and the palate and the exterior conformation of the face. He can turn small articles on the lathe, and do needlework; he levels with a square; he puts together and takes to pieces simple machines. He is so skilled in music as to play a piece when he has been told the notes and their value. He judges the duration of time much more accurately than we by the succession of actions and of thoughts. A smooth skin, firm flesh, an

elegant shape, sweet breath, charm of voice and graceful pronunciation are qualities he prizes very highly.

He married to have eyes of his own. Before this, he had an idea of
taking a deaf man as his partner, to whom he could lend ears in
exchange for eyes. I could not sufficiently wonder at his singular
address in a great many things; and on our expressing our surprise, 'I
perceive, gentlemen,' said he, 'that you are not blind: you are
astonished at what I do, and why not as much at my speaking?' There
is more philosophy, I believe, in this answer of his than he was aware
of. The facility with which we are brought to speak is not a little
surprising. We have a number of ideas which cannot be represented
by sensible objects, and which have no substance, as it were, and we
are obliged to find terms for them by making use of a number of
ingenious and profound analogies observed between them and the
ideas they suggest. Thus a blind man should find greater difficulty in
learning to speak because there is a much larger number of imperceptible objects in his world, and thus his field for comparing and
combining is much more limited. How, for example, can he rightly
use the word *expression* (of countenance)? It is the same of many
things imperceptible to the blind; and for us who see, it is often found
hard to explain very precisely what it is. If it largely resides in the eye,
touch will be useless; and what does a blind man make of dead eyes,
or sparkling or expressive eyes? I infer from this that we unquestion-
ably derive great advantages from the concurrence of our senses and
our organs; still, it would be quite another thing if we used them
separately, and never employed two when one would suffice. To add
touch to sight, when sight would do the business, is like adding to a
carriage with two stout horses a third, which will draw one way while
the others draw another.

As to me it has always been very clear that the state of our organs
and our senses has a great influence on our metaphysics and our
morality, and that those ideas which seem purely intellectual are
closely dependent on the conformation of our bodies, I put some
questions to the blind man about the virtues and vices. The first thing
I remarked was his extreme abhorrence of theft; possibly from two
reasons – firstly, the facility with which people could steal from him
unobserved, and secondly (and still more perhaps), because he could
be immediately seen were he to go about filching. Not that he is at
any loss to secure himself against that sense which he knows we have
above him, or that he is clumsy at hiding what he might steal.

Modesty he makes no great account of. If it were not for the weather, against which clothes are a protection, he would hardly understand their use; and he openly admits he cannot see why one part of the body should be hidden rather than another; and still less by what whim some of those parts should especially be singled out, which from their use and the indispositions they suffer ought rather to be kept free. Though living in an age when philosophy has rid us of a great number of prejudices, I do not think we shall ever arrive at such complete insensibility to the prerogatives of modesty as this blind man. Diogenes would have been no philosopher in his account.

As of all the external signs which raise our pity and ideas of pain the blind are affected only by cries, I have in general no high thought of their humanity. What difference is there to a blind man between a man making water and one bleeding in silence? Do not we ourselves cease to be compassionate when distance or the smallness of the objects produces on us the same effect as deprivation of sight upon the blind? Our virtues depend so much on the sensations we receive, and the degree by which we are affected by external things. I don't doubt that if it were not for the fear of punishment, many people would find it less disagreeable to kill a man at a distance at which he would appear no bigger than a swallow, than to cut an ox's throat with their own hands. We pity a horse in pain, and we make nothing of crushing an ant; and is it not by the same principle that we are moved? Madam, how different is the morality of the blind from ours? And how different would that of a deaf man be from his? And to one with an extra sense, how deficient would our morality appear – to say nothing more? Our metaphysics and theirs agree no better. How many of their principles are mere absurdities to us, and *vice versa*? Concerning this I might enter into details, which I am pretty certain would amuse you, but which certain people, who make a crime of everything, would not fail to exclaim against as profanity and infidelity, as if it were in my power to make the blind perceive things in a different way than they do. I will content myself with one observation, which everyone must allow, and that is, that the great argument for the wonders of nature falls flat upon the blind. The facility with which we create (if I may say so) new objects by means of a little glass, is something more incomprehensible to them than the stars which they have been condemned never to see. This luminous globe which moves from east to west surprises them less than a small fire which they can increase or diminish at will; and as they see matter in a more abstract manner than we do, they are less indisposed to believe that it thinks.

If a man who had had sight only for a day or two found himself in the midst of a blind people, he would either have to hold his peace or be considered a brain-sick fool. Every day he would come out with some new wonder, which would only be such to them, and which their free-thinkers would oppose tooth and nail. Would the apologists of religion not greatly avail themselves of such a stubborn disbelief, which, however just in some respects, is yet so very ill-founded? Don't think too long on this supposition; it will remind you of the persecutions undergone by those poor wretches who discovered truth in the dark ages and were rash enough to reveal it to their blind contemporaries, and found their bitterest enemies were those who from their circumstances and education would have seemed most likely to receive it willingly.

So much for the morals and metaphysics of the blind. I now pass on to less important matters, which have nevertheless lately been the chief subject of observation with regard to the blind ever since the Prussian oculist's arrival. First question: How can a man born blind form ideas of figures? By the movements of his own body and by stretching his hand in various directions, by passing his fingers continuously over an object, he gets an idea of space. If he passes his fingers along a taut thread, he obtains the idea of a straight line; if he follows the curve of a slack thread, that of a curve. In a more general sense, by repeated usage of the sense of touch, he has a memory of sensations experienced at different points; and he is capable of combining these sensations or points and forming figures. A straight line for a blind man who is not a geometrician is but the memory of a series of sensations of touch upon a taut thread; a curve, the memory of a series of tactile sensations referred to the surface of some concave or convex solid. In the ease of a geometrician, study corrects the idea of these lines by their properties which he discovers. But whether geometrician or not, the man born blind refers everything to his fingers' ends. We combine coloured points, he only palpable points, or, to speak more precisely, only such tactile sensations as he remembers. He does not go through a mental process analogous to ours; he does not create an image, for to do this it is necessary to colour a background and mark upon it points of a different colour from that background. Make these points of the same colour as the ground, and they are at once lost in it, and the figure disappears; at any rate, that is the ease in my imagination, and I suppose all imaginations are alike. When I propose to perceive in my head a straight line other than by its properties, I begin by spreading in it a white cloth, against which I set out a series of black points in the same direction. The stronger the colour of the ground and points, the clearer my perception of the points. To view in my imagination a figure of a colour resembling that of the ground, puts me to no less trouble than if out of myself and on a canvas. You see then, madam, that laws might be given for imagining with ease various objects variously coloured, but such laws are by no means calculated for one born blind. Such a man who cannot colour (and consequently cannot figure as we understand it) only remembers such sensation as one derives from touch, which he refers to different points, places and distances, and from which he composes figures. I believe that we who see never imagine any shape without colouring it, and that if we are given little balls in the dark, whose substance and colour are unknown to us, we shall immediately think of them as black or white, or some other colour; and that if we did not, we, like the blind man, should have the remembrance only of little sensations excited at our fingers' ends, and such as little round bodies may occasion. If this memory is very fleeting with us, if we have very little idea how one born blind fixes, recalls and combines the sensations of touch, it is owing to the custom we derive from our eyes of realising everything in our imagination through colours. It has happened, however, that during the agitations of a violent passion I felt a thrill run through my whole hand, and I felt the impression of the bodies I had touched some time ago revived as vividly as if they had been still present to my touch, and I realised very distinctly that the limits of sensation exactly coincided with those of these absent bodies. Although sensation by itself is indivisible, it occupies, if one may use the word, an extension in space to which the blind man is able to add and subtract mentally by enlarging or diminishing the parts affected. By this means he compares points, surfaces, and solids; and he could imagine a solid as large as this terrestrial globe, if he were to imagine his fingers' ends as large as this globe, and occupied by sensation in its length, breadth, and depth. I know of nothing which is a better proof of the reality of this internal sense than this faculty, weak in us, but strong in those born blind, of feeling or recalling the sensation of bodies when they are absent and no longer acting on us. We cannot make a blind man understand how imagination represents absent objects as present to us, but we can easily recognise in ourselves the faculty that the blind possess of feeling at one's fingers' ends an absent body. To do this, press the forefinger and thumb together, shut your eyes; separate your fingers, and immediately after this separation examine yourself and tell me if the sensation does not linger after the pressure has ceased; if, while the pressure lasted, your mind appears to be in your head rather than at the ends of your fingers, and if this pressure does not convey the nature of a surface by the space which the sensation occupies? We only distinguish the presence of external things from their picture in our imagination by the strength or weakness of the impression; and similarly, the blind only distinguish the sensation from the actual presence of an object at their fingers' ends, by the strength or weakness of that sensation.

If ever a philosopher, blind and deaf from his birth, were to construct a man after the fashion of Descartes, I can assure you, madam, that he would put the seat of the soul at the fingers' ends, for it is from these that the greater part of the sensations and all his knowledge are derived. Who would inform him that his head is the seat of his thoughts? If the labours of the imagination tire our brain, this is because the effort we make to imagine is somewhat similar to that to perceive very near or very small objects. But this would not be the case with a man blind and deaf from his birth, for the sensations which he has gathered from touch will be the world, so to speak, of all his ideas, and I should not be surprised if, after a profound meditation, his fingers were as wearied as our heads. I am not afraid that a philosopher might object to such an one that the nerves cause our sensations and that they all start from the brain. Were these two propositions fully demonstrated, which is very far from being the case, especially the former, an exposition of all the dreams of naturalists on this head would be sufficient to confirm him in his opinion.

But if the imagination of the blind man is no more than the faculty of calling to mind and combining sensations of palpable points; and of a sighted man, the faculty of combining and calling to mind visible or coloured points, the person born blind consequently perceives things in a much more abstract manner than we; and in purely speculative questions, he is perhaps less liable to be deceived. For abstraction consists in separating in thought the perceptible qualities of a body, either from one another, or from the body itself in which they are inherent; and error arises where this separation is done in a wrong way or at a wrong time – in a wrong way in metaphysical questions, or at a wrong time in applied mathematics. There is perhaps one certain method of falling into error in metaphysics, and that is not to simplify the subject under investigation sufficiently; and an infallible secret for obtaining incorrect results in applied mathematics is to suppose objects less compounded than they usually are.

There is one kind of abstraction of which so few are capable that it seems reserved for purely intellectual beings, and that is that in which everything is reduced to numerical units. We must admit that the results of this geometry would be very exact, and its formulas very comprehensive, for there are no objects, either possible or actually existent, which these simple units could not represent, by points, lines, surfaces, solids, thoughts, ideas, sensations, etc.; and if this should prove to be the foundation of Pythagoras's doctrine, he might be said to have failed in his aim, his mode of philosophising being too far above us, and too near that of the Supreme Being, who, according to the ingenious phrase of an English geometrician,[7] always geometrises in the universe.

But units pure and simple are too vague and general symbols for us. Our senses bring us back to symbols more suited to our comprehension and the conformation of our organs. We have arranged that these signs should be common property and serve, as it were, for the staple in the exchange of our ideas. We have made them for our eyes in the alphabet, and for our ears in articulate sounds; but we have none for the sense of touch, although there is a way of speaking to this sense and of obtaining its responses. For lack of this language, there is no communication between us and those born deaf, blind, and mute. They grow, but they remain in a condition of mental imbecility. Perhaps they would have ideas, if we were to communicate with them in a definite and uniform manner from their infancy; for instance, if we were to trace on their hands the same letters we trace on paper, and always associated the same meaning with them.

Is not this language, madam, as good as another? Is it not already to hand, and would you dare to say that you have never been communicated with by this method? Nothing remains but to fix it, and make its grammar and dictionaries, if it is found that the expression by the common characters of writing is too slow for the sense of touch. Knowledge has three routes by which it reaches our mind, and we keep one barricaded for lack of signs. If the two others had been neglected we should now be little better than beasts. Just as a pressure is the only sign we have to the touch, so a cry would have been the only sign to the hearing. We have to lose one sense before we realise the advantage of symbols given to the remainder, and people who have the misfortune to be born deaf, blind, and mute, or who have lost these three senses by some accident, would be delighted if there existed a clear and precise language of touch.

It is much easier to use symbols already invented than to invent them, as one is obliged to do when there are none current. What an advantage it would have been for Saunderson to find an arithmetic arranged with signs for the touch all ready to hand at the age of five, instead of having to invent it at twenty-five! This Saunderson, madam, is another blind man whose story you will be interested to hear. Wonderful stories, indeed, are told of him, and yet there is not one to which, from his achievements in literature and his skill in mathematics, we may not safely give him credit. He used the same machine for algebraical calculations and for the description of rectilinear figures.[8] You would be interested in an account of this if intelligible, and you will see my description assumes no more knowledge on your part than you actually possess, and that it would be very useful to you if you should ever want to make long calculations by touch.

Imagine a square such as you see in figure 1, divided into four equal parts by lines perpendicular to the sides, in such a way that it gives nine points, 1, 2, 3, 4, 5, 6, 7, 8, 9. Suppose this square perforated with nine holes to hold pins of two kinds, both of the same length and thickness, but one kind with a head larger than that of the other.

The large-headed pins are only placed in the centre of the square, the small-headed pins only on the sides, except in the single case of zero. Zero is marked by a large-headed pin placed in the centre of the small square which has no pin set on the sides. The figure 1 is represented by a small-headed pin, placed in the centre of the square, which has no pin set on its sides. The figure 2, by a large-headed pin placed in the centre of the square, and by a smallheaded pin placed in one of the sides at the point 3. The figure 3, by a large-headed pin placed in the centre of the square, and by a small-headed pin placed in one of its sides at the point 4. The figure 4, by a large-headed pin placed in the centre of the square, and by a small-headed pin placed in one of the sides at the point 3. The figure 5, by a largeheaded pin placed in the centre of the square, and by a small-headed pin placed in one of the sides at the point 4. The figure 6, by a large-headed pin placed in the centre of the square, and by a small-headed pin placed in one of its sides at the point 5. The figure 7, by a large-headed pin placed in the centre of the square, and by a small-headed pin placed in one of the sides at the point 6. The figure 8, by a large headed pin placed in the centre of the square, and by a small-headed pin placed in one of the sides at the point 7. The figure 9, by a large-headed pin placed in the centre of the square, and by a small-headed pin placed in one of the sides at the point 8.

This gives ten different symbols for the sense of touch, each of which corresponds to one of our ten arithmetical characters. Now imagine a board as large as you choose, divided into small squares arranged horizontally and separated by a small space one from the other, as you see in fig. 2, and you have Saunderson's instrument.

You can easily see that there is no number which cannot be expressed in the tablet, and hence no arithmetical process which cannot be carried out therein.

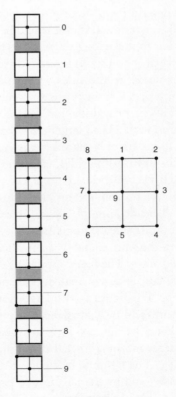

FIGURE I

Suppose, for example, that we want to find the sum of, or to add, the nine following numbers:

1	2	3	4	5
2	3	4	5	6
3	4	5	6	7
4	5	6	7	8
5	6	7	8	9
6	7	8	9	0
7	8	9	0	1
8	9	0	1	2
9	0	1	2	3

I write them on the table in the order they are named: the first figure on the left of the first number, on the first square to the left of the first line; the second figure on the left of the first number, on the

FIGURE 2

FIGURE 3

second square on the left of the same line, and so on.

I place the second number in the second row of squares; units are units, tens are tens, etc.

I place the third number in the third row of squares, and so on, as you see in fig. 2. Next, touching with my fingers each vertical row from the top to the bottom, beginning with that which is most to my left, I add together the numbers in each row expressed; and I write the tens that are over at the end of that column I pass to the second column, moving left, and work in this way; from thereon to the third, and so on completing my addition.

This is how the same tablet served him to prove the properties of rectilinear figures. Supposing he had to prove that parallelograms which have the same base and same height are equal in area, he placed his pins as you see in fig. 3; he added names to the angles, and proceeded with the proof with his fingers.

FIGURE 4

Supposing that Saunderson only used large headed pins to mark the limits of his fingers, he could arrange round these small-headed pins of nine different varieties with, all of which he was familiar with. Thus he was never at a loss, except in cases where the great number of angular points which he was obliged to name in his proof forced him to resort to the letters of the alphabet. We are not told how he used them.

We only know that his fingers moved over his tablet with astonishing rapidity; that he made the longest calculations successfully; that he could interrupt them, and recognise when he was in error; that he could verify them with ease; and that this work did not take him as much time as one might imagine, because he could arrange his tablet to suit his convenience.

This arrangement consisted in placing largeheaded pins in the centre of all the squares. This done, he had only to fix their value by

FIGURE 5

small-headed pins, except in the case when he wished to express an unit. In that case he put a small-headed pin in the centre of the square, in place of the largeheaded pin.

Sometimes, instead of forming a complete line with pins, he only placed them at all the angles or points of intersection, and round these he stretched silk threads which completed his figures. (See fig. 4)

He left several other instruments which facilitated his geometrical studies; the use he made of these is not known, and more acumen would perhaps be required to discover this than to solve a problem in integral calculus. Let a geometrician try to discover the function of four pieces of solid wood in the form of rectangular parallelepipeds, each 11 inches by 5½ wide and a little more than half an inch thick, and whose two larger opposite surfaces were divided into small squares similar to the abacus I have just described; but with this difference, that they were only perforated at certain points, in which pins were driven in up to their head. Each surface had nine small arithmetical tablets, each with ten numbers, and each of these ten numbers was composed of ten figures. Fig. 5 represents one of the small tablets, and here are the numbers it contained:

9	4	0	8	4
2	4	1	8	6
4	1	7	9	2
5	4	2	8	4
6	3	9	6	8
7	1	8	8	0
7	8	5	6	8
8	4	3	5	8
8	9	4	6	4
9	4	0	3	0

He was the author of an excellent work of its kind – *The Elements of Algebra*[9] – where the only signs of his blindness are the peculiarity of certain demonstrations which a sighted man would probably not have thought of. To him we owe the division of the cube into six equal pyramids whose apex is at the centre of the cube and the base of each is one of its faces. This is used by him as a simple proof that every pyramid is the third of a prism having the same height and the same base. His taste for mathematics, his small means, and the advice of his friends decided him to give public lectures. His marvellous facility for clear demonstration encouraged his friends to think he would prove a successful teacher, for he taught his pupils as if they could not see, and a blind man who makes himself clear to the blind must be doubly lucid to the sighted; it is a telescope the more.

His biographers say that his talk abounded in happy expressions, and I can well believe it. But 'what do you mean by happy expressions?' you will perhaps inquire. I answer, madam, it is using expressions to one sense (touch, for example) which are also metaphorical to another sense (say, sight): as a result, a double light is shed on the subject for the listener, the direct light of the natural use of the expression and the reflected light of the metaphor. It is evident that in these cases, Saunderson, with all his intelligence, was not aware of the full force of the terms he employed, since he only realised half of the ideas attached to these terms. But does this not happen to all of us at times? It may happen to idiots, who sometimes make excellent jokes, and clever folk who say a foolish thing, without either being aware of it. I have observed the want of words produces a similar effect in foreigners, who in an unfamiliar language are obliged to say everything in very few words, some of which they unknowingly use very happily. But every language being to writers of a lively imagination deficient in fit words, they are in the same case as clever foreigners: the situations invented by them, the delicate gradations they perceive in characters, the natural scenes they draw, are continually leading them away from ordinary speech and causing them to adopt turns of phrases which never fail to charm when they are neither precious nor obscure. These are faults which are more or less readily forgiven, according to the reader's wit and knowledge of the language. This is why M. de M—[10] is the French author who most pleases the English, and Tacitus, of all the classics, bears the bell among the thinkers; they do not stick to the licences of the style, it is only the truth of the expression which strikes them.

Saunderson was extremely successful as professor of mathematics at the University of Cambridge. He gave lessons in optics, he lectured on the nature of light and colours, he explained the theory of vision; he wrote on the properties of lenses, the phenomena of the rainbow, and many other subjects connected with sight and its organ.

These facts lose much of their marvellous character when you consider that there are three distinct elements in a question in which both physics and geometry enter – the phenomenon to be explained, the hypotheses of the geometrician, and the resultant calculation. Now it is manifest that, however great the penetration of the blind man, the phenomena of light and colour are unknown to him. The hypotheses he will understand, as all of them relate to palpable causes; but the geometrician's reason for preferring them to others will be out of his ken, as in order to see that he must be able to compare the hypotheses themselves with the phenomena. Therefore the blind man takes the hypotheses for what they are given him, a ray of light for a fine and elastic thread, or for a succession of minute bodies striking our eyes with incredible velocity, and he makes his calculations accordingly. The transition is made from physics to geometry, and the question becomes purely mathematical.

But what are we to think of the results of the calculation? Firstly, that it is sometimes extremely difficult to obtain them, and that it would be to little purpose that a man of science could form the most plausible hypotheses, were he not able to verify them by geometry;

accordingly the greatest physicists, Galileo, Descartes, and Newton, were great geometricians. Secondly, the results are more or less certain, as the preliminary hypotheses are more or less complex. When the calculation is based on a simple hypothesis, the conclusions have the validity of geometrical proofs. When there are a great many suppositions, the probability of each hypothesis being true diminishes in the ratio of the number of these hypotheses; but on the other hand increases owing to the improbability that so many false hypotheses could be mutually corrective and produce a result confirmed by the phenomena. A parallel to this would be an addition, of which the sum was correct although the sum of groups of numbers had been wrongly added up. We must admit that such a result is possible, but at the same time you see that it would very seldom prove so. The greater the number of numbers to be added, the greater the probability of error in the addition of each, but at the same time this probability is reduced if the result of the operation is correct. There are therefore a number of hypotheses, the certainty resulting from which would be the least possible. If I make A plus 13 plus C equal to 50, must I conclude from 50 being the real quantity of the phenomena that the suppositions represented by the letters A, B, C are true? Not at all, for there are numberless ways of subtracting from one of these letters and adding to the others which would always give 50 as the result. But the case of three combined hypotheses is perhaps one of the most disfavourable.

One advantage of calculation which I must not omit is, that the disparity found between the result and the phenomenon excludes false hypotheses. If a man of science proposes to find the curve formed by a ray of light in passing through the atmosphere, he must regulate himself by the density of the strata of air, the law of refraction, the nature and form of the luminous corpuscles, and perhaps other essential factors which he does not include in his calculation, either because he does not know them or because he deliberately leaves them out of consideration. He then determines the curvature of the ray. If the actual curve differs from that of his calculation, his hypotheses are incomplete or false. If the actual curvature agrees with that of his calculation, there are two alternatives: first that his hypotheses were mutually corrective, secondly that they were correct. But which is true? He does not know, and yet that is the certainty to which he can attain.

I read Saunderson's *Elements of Algebra* carefully in hopes of finding out what I wanted to know from those who knew him intimately, and who have related some particulars of his life; but my curiosity was baffled, and it occurred to me that elements of geometry from him would have been a work both more singular in itself and of greater use to us. We would find in it definitions of point, line, surface, solid, angle, intersections of lines, and planes, in which I make no question that he would have proceeded on principles of very abstract metaphysics, closely allied to that of the idealists. Those philosophers, madam, are termed idealists who, conscious only of their own existence and of a succession of external sensations, do not admit anything else; an extravagant system which should to my thinking have been the offspring of blindness itself; and yet, to the disgrace of the human mind and philosophy, it is the most difficult to combat, though the most absurd. It is set forth with equal candour and lucidity by Doctor Berkeley, Bishop of Cloyne, in three dialogues.[11] It were to be wished that the author of the *Essay on the Origin of Human Knowledge*[12] would take this work into examination; he would find matter for useful, agreeable, and ingenious observation – for which, in a word, no person has a better talent. Idealism deserves an attack from his hand, and this hypothesis is a double incentive to him from its singularity, and much more from the difficulty of refuting it in accordance with his principles, which are the same as those of Berkeley. According to both, and according to reason, the terms essence, matter, substance, agent, etc., of themselves convey very little light to the mind. Moreover, as the author of the *Essay on the Origin of Human Knowledge* judiciously observes, whether we go up to the heavens, or down to the deeps, we never get beyond ourselves, and it is only our own thoughts that we perceive. And this is the conclusion of Berkeley's first dialogue, and the foundation of his entire system. Would you not be curious to see a trial of strength between two enemies whose weapons are so much alike? If either got the better it would be the one who wielded these weapons with the greater address; the author of the *Essay on the Origin of Human Knowledge* has lately given in his *Treatise on Systems* additional proof of his adroitness and skill and shown himself a redoubtable foe to the systematics.

We have wandered far from the blind, you will say. True, madam, but you must be so good as to allow me all these digressions; I promised you a conversation, and I cannot keep my word without this indulgence.

I have read as carefully as was in my power what Saunderson has said on the infinite; and I assure you he had such very just and very clear notions on the subject that in his account most of our infinitarians would have been looked on as but blind. You yourself shall be judge: though this matter is somewhat difficult, and a little beyond your mathematical knowledge, I hope to bring it within your grasp and initiate you into the logic of the infinite.

The case of this famous blind man proves that the sense of touch, when trained, can become more delicate than sight, for he distinguished genuine from counterfeit coins[13] by passing his hands over a number of them, although the counterfeits were sufficiently good imitations to deceive a clearsighted connoisseur; and he judged the accuracy of a mathematical instrument by passing the tips of his fingers along its divisions. This is certainly more difficult than to judge by touch the resemblance of a bust to the person represented, and this shows that a blind people might have sculptors and put statues to the same use as among us to perpetuate the memory of great deeds, and of people dear to them; and in my opinion feeling such statues would give them a keener pleasure than we have in seeing them. What a delight to a passionate lover to draw his hand over beauties which he would know again, when illusion, which would act more potently on the blind than on those who see, should come to reanimate them! But perhaps, as he would take a deeper pleasure in the memory, his grief would be the keener for the loss of the original.

Saunderson, like the blind man of Puisaux, was affected by the smallest atmospheric change, and could recognise, especially in still weather, the presence of objects not far from him. It is related of him that being present during some astronomical observations taken in a garden, the clouds which hid the face of the sun every now and then from the spectators at the same time caused such a change in the action of the rays on his face as signified to him the moments which favoured or impeded the experiments. You may, perhaps, think that some change in the eye might indicate to him the presence of light, but not of distant objects, and I would have supposed so myself, but for the fact that Saunderson had lost not only his sight but its organ.

Saunderson, then, saw by means of his skin, and this integument of his was so keenly sensitive that with a little practice he could certainly have recognised the features of a friend traced upon his hand, and would have exclaimed, as the result of successive sensations caused by the pencil: 'That is so-and-so.' Thus the blind have likewise a painting, in which their own skin serves as the canvas. These are no wild fancies, and I am sure if the little mouth of M—— were traced on your hand, you would immediately recognise it. Yet you must admit the blind man would find this an easier task than you, accustomed though you are to see and admire that mouth. For two or three elements enter into your recognition: the comparison of the tracery on your hand with the picture formed on the ground of your eye; the recollection of the manner in which we are affected by things felt, and of the manner with which we are affected by things we have only seen and admired; finally, the application of these data to the question of the draughtsman, who asks you when he draws with his pencil a mouth on the skin of your hand: 'Whose mouth is this which I am drawing?' Whereas the sum of the sensations aroused by a mouth laid on the blind man's hand is the same as the sum of the successive sensations caused by the draughtsman's pencil.

I might add to this account of Saunderson and the blind man of Puisaux, Didymus of Alexandria, Eusebius the Asiatic, and Nicaise of Mechlin,[14] and some other people who, though lacking one sense, seemed so far above the level of the rest of mankind that the poets might without exaggeration have feigned the jealous gods to have deprived them of it, from fear that mortals should equal them. For what was Tiresias, who had penetrated the secrets of the gods, but a blind philosopher whose memory has been handed down to us by fable? But let us return to Saunderson and follow the history of this extraordinary man to his grave.

When he was at the point of death,[25] a clergyman of great ability, Mr Gervase Holmes, was summoned to his side, and they held a discussion upon the existence of God, some fragments of which are extant, and which I will translate to the best of my ability, for they are well worth it. The clergyman began by haranguing on the wonders of nature. 'Ah, sir,' replied the blind philosopher, 'don't talk to me of this magnificent spectacle, which it has never been my lot to enjoy. I have been condemned to spend my life in darkness, and you cite wonders quite out of my understanding, and which are only evidence for you and for those who see as you do. If you want to make me believe in God you must make me touch Him.' 'Sir,' returned the clergyman, very appositely, 'touch yourself, and you will recognise the Deity in the admirable mechanism of your organs.'

'Mr Holmes,' replied Saunderson, 'I must repeat it, all that does not appear so admirable to me as to you. But even if the animal mechanism were as perfect as you maintain, and I dare say it is (for you are a worthy man and would scorn to impose on me), what relation is there between such mechanism and a supremely intelligent Being? If it fills you with astonishment, that is perhaps because you are accustomed to treat as miraculous everything which strikes you as beyond your own powers. I myself have so often been an object of admiration to you, that I do not have a very high idea of your idea of the miraculous. I have had visits from people from all parts of England who could not conceive how I could work at geometry: you must allow such folk not to have been very exact in their notions of the possibility of things. We think a certain phenomenon beyond human power and we cry out at once: "It is the handiwork of a god"; our vanity will stick at nothing less. Why can we not season our talk with a little less pride and a little more philosophy? If nature offers us a knotty problem, let us leave it for what it is, without calling in to cut it the hand of a being who immediately becomes a fresh knot and harder to untie than the first. Ask an Indian how the earth hangs suspended in mid-air, and he will tell you that it is carried on the back of an elephant; and what carries the elephant? A tortoise. And the tortoise? You pity the Indian, and one might say to yourself as to him: "My good friend Mr Holmes, confess your ignorance, and drop the elephant and the tortoise.'"[16]

Saunderson paused, apparently waiting for a reply, but what possible reply was there to the blind man? Mr Holmes availed himself of his good opinion of his probity and of the abilities of Newton, Leibniz, Clarke, and some of his fellow-countrymen, men of the highest genius, who had all been impressed by the wonders of nature and recognised an intelligent being as its creator. This was certainly the clergyman's strongest argument. The blind man admitted that it would be presumptuous to deny what such a man as Newton had acquiesced in; yet he represented to the clergyman that Newton's evidence was not of that weight to him, as that of all nature to Newton; while Newton believed on God's word, he was reduced to believe on Newton's word.

'Consider, Mr Holmes,' he added, 'what a confidence I must have in your word and in Newton's. Though I see nothing, I admit there is in everything an admirable design and order. I hope you will not demand more. I take your word for the present state of the universe, and in return keep the liberty of thinking as I please on its ancient and primitive state, with relation to which you are as blind as myself. Here you will have no witnesses to confront me with, and your eyes are quite useless. Think, if you choose, that the design which strikes you so powerfully has always subsisted, but allow me my own contrary opinion, and allow me to believe that if we went back to the origin of things and scenes and perceived matter in motion and the evolution from chaos, we should meet with a number of shapeless creatures, instead of a few creatures highly organised. I make no criticism on the present state of things, but I can ask you some questions as to the past. For instance, I may ask you and Leibniz and Clarke and Newton, who told you that in the first instances of the formation of animals some were not headless and others footless? I might affirm that such a one had no stomach, another no intestines, that some which seemed to deserve a long duration from their possession of a stomach, palate, and teeth came to an end owing to some defect in the heart or lungs; that monsters mutually destroyed one another; that all the defective combinations of matter disappeared, and that those only survived whose mechanism was not defective in any important particular and who were able to support and perpetuate themselves.[17]

'Suppose the first man had his larynx closed, or had lacked suitable food, or had been defective in the organs of generation, or had failed to find a mate, or had propagated in another species, what then, Mr Holmes, would have been the fate of the human race? It would have been still merged in the general depuration* of the universe, and that proud being who calls himself man, dissolved and dispersed among the molecules of matter, would have remained perhaps for ever hidden among the number of mere possibilities. If shapeless creatures had never existed, you would not fail to assert that none will ever appear, and that I am throwing myself headlong into wild fancies, but the order is not even now so perfect as to exclude the occasional appearance of monstrosities.' Then, turning towards the clergyman, he added, 'Look at me, Mr Holmes. I have no eyes. What have we done, you and I, to God, that one of us has this organ while the other has not?'

Saunderson uttered these words in such a sincere and heartfelt tone that the clergyman and the rest of the company could not remain insensible to his suffering, and began to weep bitterly. He noticed it and said to the clergyman, 'Mr Holmes, I was aware of the kindness of your heart, and I am very grateful for the expression of it you have given me just now; but if you love me, do not grudge me my dying consolation of never having caused anyone affliction.'

---

* To Depurate: To purify; to cleanse; to free anything from its impurities (Johnson's *Dictionary*).

Then, continuing the conversation in a firmer tone, he added: 'I conjecture, then, that in the beginning, when matter in a state of ferment brought this world into being, creatures like myself were of very common occurrence. But might not worlds too be in the same case? How many faulty and incomplete worlds have been dispersed and perhaps form again, and are dispersed at every instant in remote regions of space which I cannot touch nor you behold, but where motion continues and will continue to combine masses of matter, until they have found some arrangement in which they may finally persevere? O philosophers, travel with me to the confines of this universe, beyond the point where I feel and you behold organised beings; cast your eyes over this new ocean, and search in its aimless and lawless agitations for vestiges of that intelligent Being whose wisdom fills you with such wonder and admiration here!

'But what is the use of taking you out of your element? What is this world, Mr Holmes, but a complex, subject to cycles of change, all of which show a continual tendency to destruction; a rapid succession of beings that appear one by one, flourish and disappear; a merely transitory symmetry and momentary appearance of order? A moment ago I reproached you for estimating the perfection of things by your own capacity; I might accuse you here of measuring duration by your own existence. You judge the phases of the world's existence to be like the ephemeral insect of yours. The world seems to you eternal, just as you seem eternal to creatures which last only a day; and the insect is more reasonable than you. What a prodigious series of ephemeral generations witness to *your* eternity, what an immense tradition! Yet we shall all pass away without a possibility of denoting the real extent which we took up, or the precise time of our duration. Time, matter, and space are perhaps but a point.'

During this conversation Saunderson became more excited than his state of health would permit, and an attack of delirium ensued, which lasted several hours. At its close he cried, 'O thou God of Clarke and Newton, have mercy on me!' and expired.

Such was the end of Saunderson. You see, madam, that all the arguments of the clergyman he took exception to were not the sort to convince a blind man. What a disgrace to men who have no better argument than him; men who have eyes, to whom the marvellous spectacle of nature from sunrise to the setting of the smallest stars reveals the existence and glory of its Maker! They have sight, which Saunderson was deprived of, but Saunderson was blessed with a purity of life and uprightness which we look for in vain in them. Accordingly they lead the life of the blind, and Saunderson died as if he knew the light. The voice of nature made itself clear to him through the senses he possessed, and his evidence is more convincing against those who obstinately shut their eyes and ears. Was the true God not more completely veiled by the mists of paganism for a Socrates, than for the blind Saunderson, who never enjoyed the spectacle of nature?

I am very sorry, madam, both for your sake and mine, that no further interesting particulars of this famous blind man have been handed down. His conversation would perhaps have afforded more light than all our experiments. Those about him must have been devoid of the philosophic spirit. I make an exception in favour of his pupil, Mr William Inchlif, who only saw Saunderson during his last moments, and who took down his last words, which I should advise all who know English to read in the original, printed in Dublin in 1747, and entitled *The Life and Character of Dr Nicholas Saunderson, late Lucasian Professor of the Mathematicks in the University of Cambridge; by his disciple and friend William Inchlif, Esq.*[18] They will find a charm, and a vigour in this, seldom rivalled, but which I do not flatter myself I have conveyed in translation, in spite of all my care.

He married in 1713 the daughter of Mr Dickons, rector of Boxworth, in the county of Cambridge, and had by her a son and daughter who are still living. His farewell to his family was exceedingly moving. 'I go,' said he, 'to our common destination; spare me laments which unman and unnerve. The expressions of grief which escape you only make me conscious of my own. I gladly give up a life which has been for me a long desire, a constant privation. Live on, as virtuous as I, but more fortunate, and learn to die with equal calm.' He then took his wife's hand, which be held for a moment clasped in his own; he turned his face towards her as if he desired to see her; he blessed his children, embraced them, and begged them to leave him, because they caused him greater grief than the approach of death.

England is the land of philosophy and of scientific inquiry, yet without Mr Inchlif we should only know what the common man could have told of Saunderson; for instance, that he recognised places he had once visited by the sound the walls and pavement reflected, and many similar anecdotes, all equally common to the majority of the blind. Strange! Are blind men of such high intellectual abilities as Saunderson of common occurrence in England, and are men born blind who lecture on optics to be found every day?

People try to give those born blind the gift of sight, but, rightly considered, science would be equally advanced by questioning a sensible blind man. We should learn to understand his psychology and should compare it with ours, and perhaps we should thereby come to a solution of the difficulties which make the theory of vision and of the senses so intricate and so confused. But I own I cannot conceive what information we could expect from a man who had just undergone a painful operation upon a very delicate organ which is deranged by the smallest accident and which when sound is a very untrustworthy guide to those who have for a long time enjoyed its use. For my part, as to the theory of the senses, I would sooner hear a metaphysician who was acquainted with the principles of metaphysics, the elements of mathematics, and the conformation of the organs of sense, than an uneducated man whose sight was first due to an operation for cataract. I would have less confidence in the impressions of a person seeing for the first time than in the discoveries of a philosopher who had profoundly meditated on the subject in the dark; or, to adopt the language of the poets, who had put out his eyes in order to be the better acquainted with vision.

To obtain some certainty in such experiments the subject must at least have been prepared a long time beforehand; he should be made a philosopher – no rapid process even with a philosopher for teacher! And imagine the task if the teacher were not enlightened, or (worse still) fondly and mistakenly imagined himself enlightened! It would be better to postpone the investigation to a considerable period after the operation. To do this, the patient would have to remain in darkness, and the investigator would have to see to it that his wound was healed and his eyes perfectly sound. I would not expose him to full daylight for the first time. A strong light dazzles our eyes; what effect will it not have on an organ which must be extremely tender and sensitive, and which has never yet felt any impression to blunt it?

But this is only the beginning. It would be a difficult and delicate task to reap any benefit even from a person thus prepared, and to adapt our questions so that he may precisely say only what passes in himself. This interrogatory should be held in presence of the Academy; or rather, to avoid the presence of idle spectators, only such as deserve that distinction by their knowledge of philosophy, anatomy, etc., should be invited.

The task would not be beneath the intelligence of the best and wisest of men; to train and question one born blind would be an occupation worthy of the combined talents of Newton, Descartes, Locke and Leibniz.

I will end my letter, which I admit is already too lengthy, by a problem which was propounded some time ago. Some reflections upon Saunderson's singular condition tend to show that it has never been absolutely solved. Suppose one blind from birth has been taught to distinguish by touch a cube and a sphere of the same metal and of approximately the same size, so that when he touches them he can say which is the cube and which is the sphere. Suppose the cube and sphere placed on a table and the blind man suddenly to see; can he distinguish the cube from the sphere by sight without touch?

Mr Molyneux first stated this problem and attempted to solve it. He declared that the blind man would not distinguish between the cube and the sphere; 'for,' said he, 'though he has learnt by experience the effect of a sphere and a cube upon the sense of touch, he does not yet know that what affects his sense of touch in such and such a manner must affect his sight thus or thus; nor that the projecting angle of the cube which presses against his hand should appear to his eyes as it actually does appear in the cube.'

Locke,[19] when consulted on this point, said: 'I certainly agree with Mr Molyneux's opinion. I believe the blind man incapable at first sight of affirming with any certainty which was the cube and which the sphere if he merely looked at them, although, if he touched them, he could name them and distinguish between them by the difference of their shape, which he would recognise by touch.'

The Abbé de Condillac,[20] whose *Essay on the Origin of Human Knowledge* you have read with so much pleasure and profit, and whose excellent *Treatise on Systems* accompanies this letter, makes an original contribution to the question. I shall not repeat his arguments here, since you will have the pleasure of reading his book in which they are expounded in such an entertaining and yet such a philosophical manner that it would be a mistake on my part to tear them from their context. I shall merely observe that they all tend to prove that the born-blind either sees nothing, or distinguishes between the sphere and the cube; and that the conditions that these two bodies should be of the same metal and of approximately the same size (which was postulated in the problem) are unnecessary, which cannot be disputed; for he might have said, if there be no essential connection between the sensations of the sight and the touch (as Messrs Locke and Molyneux assert), they must admit that a body may to the eye appear to have two feet in diameter which yet would vanish on being touched. De Condillac adds, however, that if the blind man sees bodies and distinguishes their forms, and yet hesitates what to think about them, it must be from metaphysical reasons, and those not a little subtle, which I shall presently explain. We have here two different opinions on the same question – a difference between philosophers of the highest rank. One would suppose, after the problem had been studied by men such as Messrs Molyneux, Locke and the Abbé de Condillac, that nothing more could be said; but the same thing can be viewed from so many different sides that it is not strange if they have not exhausted all its possibilities.

Those who declare that a man blind from birth could not distinguish between a cube and a sphere have set out by assuming a fact which perhaps should have been investigated; that is, whether a blind man who has had his cataracts removed is in a condition to use his eyes immediately after the operation. They merely say: 'The blind man, comparing the ideas of spheres and cubes which he has received by the sense of touch with those received by sight, will necessarily know them to be the same; and it would be indeed odd if he were to name that body a cube which gives the eye the idea of a sphere, and sphere that which gives the idea of a cube. He will therefore call those bodies spheres and cubes at sight which he called spheres and cubes by the sense of touch.'

But how do their antagonists reply? They have also taken for granted that the blind man could see immediately his organ was perfect; they supposed that an eye couched for cataract was like an arm that ceases to be paralysed. As the latter does not need exercise before it feels, they said, neither does the former before it sees; and they added: 'Let us grant the blind man a little more philosophy than you afford him, and after carrying on the reasoning where you left it, he will continue thus: 'But still, who is to assure me that when I approach these bodies and touch them with my hands they will not on a sudden deceive my expectation, and that a cube will not give me the sensation of a sphere and a sphere of a cube? Experience alone can teach me whether there be an analogy between sight and touch. The reports of these two senses may well be contradictory without my knowing it; no, I should perhaps suppose what is actually present to the sight to be only a mere appearance, had I not been informed that they are the very same bodies I had touched. This object certainly seems to be the body which I called a cube; and that, the body I called a sphere; but the question is, not what I *think*, but what *is*; and I am not in a position to answer the latter question satisfactorily.'"

The line of argument, says the author of the *Essay on the Origin of Human Knowledge,* would be extremely perplexing to someone who had been born blind, and I see nothing but experience which can furnish an answer to it. It seems probable that the Abbé de Condillac means only the experiment repeated by the blind man himself on a second handling of these bodies. You will soon see why I make this point. That able metaphysician might have added that the blind man would be the more inclined to suppose that two senses might be mutually contradictory, as he conceives that a mirror makes them mutually contradictory, as I have noticed already.

De Condillac proceeds to observe that Molyneux has confused the issues of the problem by laying down several conditions which are irrelevant to the metaphysical difficulties which the blind man would experience. This criticism is the more just, as the supposition of the blind man being acquainted with metaphysics is not at all out of the way; because the experiment in all such philosophical questions should be accounted to be made on a philosopher – that is to say, on a person who perceives in the questions propounded all that his reason and the state of his organs permit him to perceive. Such, briefly, are, madam, the pros and cons of the problem; and you shall now see by my examination of it how very far they, who asserted that the blind man would see geometrical figures and distinguish between them, were from realising that they were right; and what good reason their opponents had to think that they were not in the wrong.

This problem of the blind man, stated in somewhat more general terms than by Molyneux, embraces two problems which we will consider separately We may ask (1) if the blind man would see immediately after the operation for cataract; (2) supposing he is able to see, could he see well enough to distinguish between figures; could he, in seeing them, correctly give them the same names which he gave them by the sense of touch; and if he can, prove that these names are the right ones?

Will the man born blind see immediately after the cure of the organ? Those who maintain that he will not see, say: 'As soon as the blind man is able to use his eyes, all the scene before him is represented at the back of the eye. This image, which is composed of a number of objects concentrated in a very small space, is just a confused mass of figures which he will not be able to distinguish. People are on the whole agreed that it is only experience which can enable him to judge the distance of objects, and that he is obliged to approach, touch, draw back from, and again approach and touch them to assure himself that they are not part of himself and are foreign to his essence; that he is now near and now far from them. Why should experience not be a necessary preliminary for perceiving them? Without previous experience he who perceives objects for the first time would suppose, when he is out of sight of them, that they had ceased to exist; for it is only our experience of permanent objects and such as we find again in the same place where we left them which proves the continuity of their existence when out of our sight. It is perhaps for this reason that children are so readily consoled for toys taken from them. It cannot be said that they promptly forget them, for some children only two and a half years old know a considerable number of words of a language and are more at a loss to pronounce them than to retain them. Now, this is a proof of childhood's being the very season of memory.

Is it not a more likely hypothesis that children think that what they no longer see no longer exists, especially as their joy when things they have lost sight of appear again is mixed with surprise? Nurses help them to acquire the notion of the continuance of absent persons by playing a game which consists in hiding the face, and showing it again. Thus they learn a hundred times in a quarter of an hour that what ceases to appear does not necessarily cease to exist. From this it follows that we owe the notion of the continuous existence of objects to experience, of their distance to the sense of touch; that the eye may perhaps have to learn to see as the tongue to speak; that it would not be surprising should the aid of one of the senses be necessary to another; and that touch, which ascertains the existence of objects exterior to ourselves when present to our eyes, is similarly the sense to which the confirmation not only of their figures, and other details of these objects, but even their presence, is reserved.

To these arguments may be added the famous experiment of Cheselden.[21] The young man from whose eyes this skilful surgeon removed cataracts was for a long time unable to distinguish dimensions, distances, positions, or even figures. An object an inch in size held before his eye so as to hide a house from him appeared as large as the house itself. All he saw seemed as close to his eye as the object he touched to the skin. He could not distinguish what he judged round by touch from what he had judged angular; nor distinguish by sight whether what he had felt to be above or beneath him were in reality above or beneath him. He eventually succeeded, but not without difficulty, in perceiving that his house was larger than his room, but he could not conceive how this could be ascertained by sight. Repeated experiments were necessary before he became assured of paintings representing solid bodies; and when he was quite convinced by looking at pictures that what he saw was not bare surfaces, on putting his hand to a picture he was vastly surprised at finding a plane surface without any relief. He then asked which was deceptive, the sense of touch or the sense of sight? Painting likewise has the same effect on savages. They take the painted figures for living men, question them and are astonished at receiving no answer; and this error in them certainly did not proceed from their not being accustomed to see.

But what can be said about the other difficulties? That the trained and practised eye of a man sees better than the weak and untrained organ of an infant, or of one born blind who has had his eyes couched. Look, madam, at the proofs adduced by the Abbé de Condillac at the end of his *Essay on the Origin of Human Knowledge,* where he also adduces Cheselden's experiments as related by Voltaire. The effects of light upon an eye for the first time so affected, and the conditions required in humours* of that organ, the cornea, the crystalline lens, etc., are clearly and ably specified there, and leave little doubt that the vision of an infant opening its eyes for the first time, or a blind person who has just been operated upon, is very imperfect.

We must, therefore, admit we perceive a multitude of details in objects unperceived by the infant or one born blind, though these objects are equally represented at the back of their eyes; for objects to strike us is not enough – we must further attend to these experiences; that, consequently, we see nothing the first time we use our eyes; and during the first moments of sight we only receive a mass of confused sensations, which are only disentangled after a time and by a process of reflection. It is by experience alone that we learn to compare our sensations with what occasions them; that sensations having no essential resemblance with their objects, it is from experience that we are to inform ourselves concerning analogies which seem to be merely positive. In short, that touch is of great service in giving the eye an accurate knowledge of the conformity of the object to the sense-impression received of it is unquestionable; and I am much inclined to think that were everything in nature not subject to infinitely general laws – if, for instance, the pricking of certain hard bodies were

* Aqueous or vitreous humour: transparent substance before (or behind) the lens of the eye (OED).

painful, and that of certain other bodies pleasurable – we should die before we had received the hundred-millionth fraction of the experiences necessary for the preservation of our body and our well-being.

I am not, however, of opinion that the eye is incapable of learning, or, if I may say so, of experimenting alone. To ascertain the existence and form of objects by touch, there is no need to see; why should touch be necessary for complete realisation of the same objects by sight? I am awake to all the advantages of touch; I have not disguised them in these observations on Saunderson or the blind man of Puisaux; but I cannot allow it that prerogative. It is easy to see that the use of one sense may be perfected and accelerated by the observations of another; but not that there is an essential interdependence between their functions. There exist certainly properties in bodies which we should never perceive without touch; and by touch we learn the presence of certain details invisible to the eye, which only becomes aware of these when informed by the sense of touch; but their services are mutual; and in the case of persons who have sight more highly developed than touch it is the former which warns the latter of the existence of objects and of details which would pass unnoticed because of their minuteness. If unknown to you a piece of paper or some smooth, thin, and flexible substance were placed between your thumb and index finger, it is your eye alone which would inform you that the contact between your two fingers was not direct. It would be much more difficult, I may cursorily add, to deceive a blind man than a person used to see in this.

An eye which is in sound condition and freely exercised might have some difficulty in convincing itself that exterior objects are not part of itself; that some things are distant, some near; that they have forms; that some are larger than others; that they have depth, etc.; still, I make no doubt that at length it would come to see them, and to see them so distinctly as to distinguish at least their more obvious limits.

To deny this would be to set aside the aim and object of the organs; it would be forgetting the chief phenomena of vision; it would be concealing from oneself that there is no painter of such skill as to rival the beauty and exactness of the miniatures which are painted in the back of your eyes; that there is nothing more exact than the likeness of the representation to the object itself; that the canvas of this picture is not so very small, that there is no confusion among the various forms, and that they occupy about a square half-inch; and that nothing is more difficult to explain than how the sense of touch would begin to teach the eye to see were the use of the latter organ absolutely impossible without the aid of the former.

But, instead of bare presumptions, I ask you whether it is touch that teaches the eye to distinguish colours? I do not suppose such an extraordinary claim will be made for touch; and this being so, it follows that if a blind man who has just been given the gift of sight is shown a black cube or a red sphere on a white background, he will immediately discern the several outlines of these figures.

Delay will be caused, some may object, by the time which must elapse for the humours of the eye to assume their proper dispositions, for the cornea to assume the convexity requisite for vision, for the pupil to be susceptible of the dilation and contraction proper to it, for the filaments of the retina to be sensitive in the right degree to the action of light, for the crystalline to exercise its forward and backward movement or for the muscles to fulfil their functions well, for the optic nerves to become accustomed to the transmission of sensation, for the entire eyeball to accommodate itself to all the necessary dispositions, and for all its component parts to combine in the execution of that miniature, which so much illustrates the demonstration that the eye will bring itself to the requisite experience.

I admit that, plain as the picture is which I have now represented to the eye of one born blind, he will not be able clearly to distinguish its parts until all these above conditions are combined; but that is perhaps the work of a moment; and it would not be difficult, by applying the aforesaid argument to a complicated mechanism such as a watch, to prove by enumerating all the movements which take place in the drum, the fusee,* the wheels, the pallets, the pendulum, etc., that the hand would take a fortnight in moving the space of a second. If it is objected that these movements are simultaneous, I reply that so perhaps are the movements in the eye when it opens for the first time, and most of the consecutive judgments. Whatever is necessary in the eye for vision, it must be granted that it is not touch which imparts them to it, that the organ acquires them independently; consequently, will succeed in distinguishing the figures represented there without the aid of another sense.

But when does this occur?, some will say. Perhaps far sooner than is thought. When we went together to the Jardin Royal, do you

---

* Fusee: the core round which is wound the cord or chain of a clock or watch (Johnson's *Dictionary*).

remember the experiment with the concave mirror and your fright when you saw the point of a sword making at you with the same swiftness as the point of that which you pushed towards the surface of the mirror? And yet you were sufficiently accustomed to refer objects represented in mirrors to something beyond them. Experience, therefore, is not so very necessary, nor so infallible as imagined, for perceiving objects or their images where they are. Your very parrot gives proof of it. The first time he saw himself in a mirror, he touched it with his beak, and as he did not reach himself (whom he took for a fellow-parrot) he walked round the mirror. I am not for laying more than due weight on the instance of the parrot; still, it is an experiment with an animal in which preconceived notions cannot be supposed to have any share.

Yet if I were told that a man born blind saw nothing for the space of two months, I should not be surprised. I shall only conclude from it the necessity of the organs becoming practised, not the necessity of touch. It will be another reason why it is important to let such a person remain for some time in the dark, when he is to be the subject of experiment; to allow him the opportunity of exercising his eye, which will be done more conveniently in the dark than in full daylight; and only to permit a kind of twilight during the experiments, or at least to arrange for the increasing or diminishing of light at pleasure in the spot where the experiments take place. I shall only be the more inclined to agree that such experiments will always be very difficult and uncertain; and that the best and shortest (though superficially the longest) way would be to arm the subject with a philosophical training sufficient to enable him to compare the two conditions he has known, and to acquaint us with the difference between the state of a blind person and of one who has his sight. Once more, what precision is to be expected from one who has not the habit of thought and analysis, and who, like Cheselden's blind man, is so ignorant of the benefits of sight as to be insensible to his misfortune, not conceiving that the lack of this sense very much impairs his pleasure? Saunderson, who certainly deserves the name of philosopher, was not thus indifferent, and I doubt much whether he would have agreed with the author of the excellent *Treatise on Systems*; I suspect the latter to have fallen into a 'system' himself when he writes that, 'had the life of man been only an uninterrupted sensation of pleasure or of pain, happy without prospect of pain, wretched without any prospect of pleasure, he would have rejoiced or suffered; and that if he were so constituted, he would not have looked about him to discover if some influence were well disposed towards him, or desired to injure him; it is only the alternation between these two conditions which causes him to reflect,' etc.

Can you believe, madam, that by a clear train of reasoning (for that is the author's method of philosophising) he would ever have been led to this conclusion? Happiness and misery are not the same as light and darkness; the one is not simply the privation of the other. We might, perhaps, have entertained the idea that happiness was as essential to us as existence and thought, had we enjoyed it without intermission; but I cannot say the same with regard to unhappiness. It would have been very natural to look on it as a forced condition, to feel oneself innocent, yet to believe oneself guilty and to accuse or excuse nature as at present.

Does the Abbé de Condillac suppose that a child in pain only cries from his pain not having been without intermission from his birth? If he replies that 'existence and pain would be one and indivisible for one who had always suffered, and that such an one could not imagine cessation of suffering without cessation of his existence,' I make reply: 'The man living in continual misery possibly might not have said, "What have I done that I should suffer in this way?" but why might he not have said, "What have I done that I should be brought into being?"' At the same time, I see no reason why he should not have used his two synonymous verbs, I *exist* and I *suffer*, the one in prose, the other in poetry, as we use the two expressions, I *live* and I *breathe*. Moreover, madam, you will observe better than I do that this passage of the Abbé de Condillac's is admirably fine, and I fear you may say, after comparing my criticism with his reflections, you prefer an error of Montaigne's to a truth of Charron's.

You may blame my continual digressions. But digressions are of the essence of this treatise. Now my opinion on the two foregoing questions is this: the first time the eyes of one born blind open to the light, he will see nothing at all; some time will be necessary for his eye to practise sight; it will practise alone and without the aid of touch, and will eventually distinguish not only colours but the main outlines of objects. Supposing he acquired this aptitude in a very short space of time, or acquired it by using his eyes in the dark apartment in which he had been confined and urged to use that exercise for some time after the operation and before the experiments; let us now see whether he would recognise at sight the bodies he had touched enough to give them the proper names. This is the final question.

In order to treat the question in the manner you will appreciate – for you like method – I will classify the people on whom the experiment might be made. If they are dullards without education and knowledge and also unprepared, I hold that when the operation for cataract has completely removed the defect of the eye and the eye is in a healthy state, objects would be very distinctly pictured in it; but such patients, being unaccustomed to any kind of reasoning and not knowing anything of sensation or idea, would be unable to compare the sensations they had received by touch with those they now receive by sight, and would at once exclaim, 'There is a round, there is a square,' so that their judgment is not to be relied on; or even they will possibly own that they saw nothing in the objects present to their sight like what they have handled.

There is another class, who by comparing the forms they see with the bodies that had previously made an impression upon their hands, and mentally applying touch to distant objects, would describe one body as a square, and another as a circle without well knowing why, their comparison of the ideas they have acquired by sight not being sufficiently distinct in their minds to convince their judgment.

I pass to a third class of subject, a metaphysician. He, I make no doubt, would, as soon as he began to see objects clearly, reason as if he had seen these bodies all his life; and after comparing the ideas acquired by sight with those acquired by touch he would declare as confidently as you or I: 'I am very much inclined to think that this is the body which I have always called a circle, and that again what I named a square, but will not assert it to be really so. What is to prevent their disappearance if I were to touch them? How am I to know whether the bodies I see are also meant to be touched? I do not know whether visible things are palpable; but were I assured of this, and if I took the word of those about me that what I see is really what I have touched, I would be no better off. These bodies may transform themselves in my hands and transmit on contact sensations quite different from those resulting from sight. 'Gentlemen,' would he conclude, 'this body appears to be the square, that the circle; but that they are the same to touch as to sight is what I have no knowledge of.'

If we take as our subject a geometrician instead of a metaphysician, he will similarly say of the two figures he has before his eyes, one is what he used to call a square, the other what he used to call a circle: 'For I see,' he would add, 'that it is only in the former I could arrange my threads and insert my large-headed pins which denoted the angles of the square; and only in the latter figure I could place the threads I required to demonstrate the properties of a circle. Here is a circle, then, and here is a square. But,' he would have added with Locke, 'perhaps when I lay my hands on these figures they will change one into another, so that the same figure would serve me in demonstrating the properties of a circle to the blind and the properties of a square to the sighted. I might possibly see a square and at the same time feel a circle. No,' he would have continued, 'I am wrong. Those to whom I demonstrated the properties of the circle and the square did not have their hands on my abacus, and did not touch the threads which I had stretched to outline my figures, and yet they understood me; they therefore did not see a square when I felt a circle, otherwise we should have been at cross-purposes; I should have been outlining one figure and demonstrating the properties of another, I should have given them a straight line for the arc of a circle, and an arc for a straight line: but as they all understood me, all men see alike: what they saw as a square, I see as a square; what they saw as a circle, I see as a circle. So this is what I have always called a square and that is what I have always called a circle.' I have substituted a circle for a sphere and a square for a cube, because there is reason to think that we only judge of distances by experience; and of course he who uses his eyes for the first time sees only surfaces without knowing anything of projection, since a projection consists in certain points appearing nearer to us than others.

But even if the blind man were able in his first attempt to judge the projection of solidity of bodies and distinguish not only a circle from a square but likewise a sphere from a cube, yet I do not therefore think that this will hold good with regard to the case of more composite bodies. There is reason to suppose that Monsieur de Reaumur's blind woman distinguished between colours, but the odds are thirty to one that what she said of the sphere and the cube was purely guesswork. I am firmly persuaded that it was not possible for her (without inspiration) to recognise her gloves, her dressing-gown, and her shoes. These objects are so composite and full of detail; there is so little resemblance between their total shape and that of the limbs they are designed to adorn or cover that Saunderson would have been infinitely more perplexed to find out the use of his mortar-board than d'Alembert or Clairaut to discover the use of his tables.

Saunderson would infallibly have supposed a geometrical relation between the object and its use, hence he would have recognised that his skull-cap was made for his head, for this had no arbitrary form to confuse him. But what would he have thought of the points and tassel of his mortarboard? What was the use of the tassel, or why four points rather than six? And these two ornamental peculiarities would for him have been the source of a number of absurd theories, or rather an excellent satire upon what we call good taste.

Taking everything into mature consideration, it will be admitted that the difference between a person who has always seen, but to whom the use of an object is unknown, and one who knows the use of an object, but has never seen, is not to the latter's advantage. Yet, do you think, madam, if you were shown a head-dress to-day for the first time, you would ever guess it to be an ornament, and particularly intended for the head? But if it is more difficult for one born blind and seeing for the first time to form a correct idea of complex objects, what is there to prevent him taking a person dressed and sitting motionless in an armchair for a machine or a piece of furniture, and a tree with its leaves and branches tossed by the wind for a self-moving, animated, and thinking being? How much our senses suggest to us; and were it not for our eyes how apt should we be to suppose that a block of marble thinks and feels!

It is certain, therefore, that Saunderson would have been assured of his not being mistaken in the judgment he had just given of the circle and the square, and that there are cases when the reasoning and experience of others are of value in elucidating the relation of sight to touch, and in teaching that what a thing is to the eye, so it is to touch.

It would, however, be not the less essential in demonstrating some proposition of universal application (as it is termed) to test the proof by depriving it of the evidence of the senses; for you are very well aware, madam, that if somebody attempted to prove to you that two parallel lines seen in perspective are to be represented in a picture by two converging lines, because the two sides of an avenue appear to converge, it would be forgetting that the proposition is as true for one that is blind as for himself. But the foregoing supposition of one born blind suggests two others: firstly, of a man who had always seen, but was devoid of the sense of touch; secondly, of a man in whom the senses of sight and touch were mutually contradictory. We might ask the former whether, if the missing sense were given him, or sight were obscured by a bandage, he would recognise bodies by touch. It is clear that geometry (provided he were acquainted with that science) would be an infallible guide as to whether the evidence of the two senses were contradictory or no. All he would have to do would be to take the cube or sphere in his hand, and demonstrate its properties, and pronounce that what he feels a cube is a cube to the eye; hence it is a cube he holds. As to one who is ignorant of this science, I believe he would not more easily distinguish a cube from a sphere by touch than Molyneux' blind man distinguished them by sight.

In the case of a man in whom the sensations of sight and touch are in a perpetual contradiction, I do not know what he would think of shapes, order, symmetry, beauty, ugliness, etc. In all probability he would relate to those things as we relate to the real extension and real duration of beings. He would, in general, say that a body possesses a shape, but he must be inclined to think that it is neither that which he sees nor that which he feels. Such a one might be dissatisfied with his senses, but his senses would be neither satisfied nor dissatisfied with the objects. Were he inclined to charge one sense with inaccuracy, I imagine it would be touch. A hundred circumstances would incline him to think that the form of objects changes rather by the action of his hands upon them than by that of the objects on his eyes. But in consequence of these preconceived notions, the difference between hardness and softness which he would find in bodies would be very perplexing to him.

But does it follow that figures are better known to us because our senses are not self-contradictory? Who has told us that they are not false witnesses? Yet we pass judgment. Alas! madam, when we weigh our human knowledge in Montaigne's scale, we are almost reduced to adopting his motto. For what do we know? What of the nature of matter? Nothing. What of the nature of spirit and thought? Still less. What of the nature of movement, space and duration? Absolutely nothing. What of the truths of geometry? Ask any honest mathematicians, and they will admit to you that all their propositions are identical, and that so many volumes upon the circle (for example) are nothing but repetitions by a hundred different methods that it is a figure where all the lines drawn from the centre to the circumference are equal. Thus we scarce know anything, yet what numbers of books there are whose authors have all pretended to knowledge! I cannot think why the world is not tired of reading so much and learning nothing, unless it is for the very same reason that I have been talking to you for two hours, without being tired and without telling you anything.[21]

With profound respect,
I am, madam,
Your very humble and obedient servant.[22]

Addition to the Preceding Letter[23]

I am going to jot down, anyhow, on paper, certain phenomena of which I was then ignorant, and which will serve as proofs or refutations of certain paragraphs in my *Letter on the Blind*. I wrote the latter thirty-three or thirty-four years ago, and I have reread it without partiality, and am not entirely dissatisfied with it. Although the first portion seemed to me more interesting than the second, and I felt that the former could have been further extended, the latter much abbreviated, I left both as I had written them, for fear that the young man's work might suffer by the old man's retouching. I think I would find it impossible to-day to emulate all that passes muster in ideas and in expression; and I fear I am equally unable to correct what merits criticism. A famous contemporary painter spends the evening of his life in spoiling the master-pieces produced in his maturity. I do not know if the defects he finds in them are real; but either he never possessed the talent to improve them if he carried the imitation of nature to the extreme limits of art; or, if he possessed it, he has lost it, for all human qualities perish as a man decays. There comes a time when taste gives counsels which are recognised as just, but which we are unable to follow.

It is the weakness of spirit arising from the knowledge of weakness, or laziness which is one of the results of weakness and lack of spirit, which stands in the way of a labour which would detract from the value of my work rather than improve it:

> *Solve senescentem mature sanus equum, ne*
> *peccet ad extremum ridendus, et ilia ducat.[1]*
> Horace, *Epistolar.*, lib. i, epist. I, v. 8, 9.

Phenomena

I. An artist who is both an enlightened student of the theory of his art, and unequalled in its practice, has assured me that it was by touch and not by sight that he judged the soundness of kernels; and that he rolled them gently between his thumb and first finger, and so discovered by successive impressions small inequalities of surface which were invisible to his eye.

II. I have heard of a blind man who recognised by touch the colour of things.

III. I could name one who arranges the colours of bouquets with the taste upon which Jean Jacques Rousseau prided himself when, whether in jest or earnest, he confided to his friends his scheme for setting up a school to teach the flower-sellers of Paris.

IV. At Amiens there was a blind dresser who presided over numerous workmen as well as if he had the gift of sight.

V. In the case of one sighted man the use of his eyes destroyed his certainty of touch; and in order to cut his hair, he removed the mirror and placed himself before a bare wall. The blind man who does not see a danger that threatens him is the more courageous, and I am sure he would walk with firmer step over the narrow and elastic planks bridging a precipice. There are very few who are undismayed by the sight of abysses beneath them.

VI. Everyone has heard of the famous surgeon Daviel.[24] I was often present during his operations. He removed a cataract from the eyes of a blacksmith who had contracted this disease from exposure to the fire of his forge; and during twenty-five years of blindness he had grown so accustomed to the guidance of touch that he had to be forced to use the sense which had been restored to him. Daviel would beat him and say, 'Use your eyes, you wretch!' He walked and moved, and did all that we do with our eyes open, with his eyes shut.

We are drawn to the conclusion that the eye is not so necessary nor so essential to our happiness as we are inclined to believe. If the spectacle of nature had no charms for the blind smith Daviel operated on, what object is there to the loss of which we should be other than indifferent, after long deprivation of sight accompanied by no pain? The sight of a beloved woman? I don't believe it, in spite of the story I am going to relate. We imagine that if one had passed a long time without seeing, one would never be weary of looking; but that is not the case. What a contrast between momentary and constant blindness!

VII. Poor patients seeking Daviel's help were drawn to his laboratory from all the provinces of the kingdom by his charity, and his reputation also gathered there a large body of interested and learned spectators. I believe Marmontel and I were present on the same day. The patient was seated and his cataract removed; Daviel laid his hand upon the eyes which he had just restored to the light. An old woman, standing beside him, showed the liveliest interest in the success of the operation; she shook in every limb at each movement of the operator. The latter signed to her to draw near, placed her kneeling opposite the patient, and removed his hands. The patient opened his eyes, saw, and cried: 'Oh, it is my mother!' I have never heard a more piteous cry; I seem to hear it still. The old woman fainted, the spectators wept, and gave their money freely.

VIII. The most astonishing case of all those who have lost their sight almost from their infancy was Mademoiselle Melanie de Salignac, a relative of Monsieur de la Fargue, a lieutenant-general in the army, who recently died at the age of ninety-one, covered with wounds and honours. She was the daughter of Madame de Blacy, who is still living, and who never ceases to regret a child who was the delight of her life and the admiration of all her acquaintances. Madame de Blacy is a woman of high character, who is willing to confirm the truth of my account. I write from her dictation such details of the life of Mademoiselle de Salignac as did not come under my personal observation during a friendship which began with her and her family in 1760, and which lasted until 1763, the year of her death.

She had a sound judgment and great sweetness of disposition and subtlety of mind, as well as naïveté and freshness. When an aunt asked her mother to help her entertain nineteen bores at dinner, she replied, 'I do not understand my dear aunt: why be kind to nineteen bores? I only wish to be kind to those I love.'

The sound of voices had the same attraction or antipathy for her as facial expression for those who see. A relation of hers, a receiver-general of finance, unexpectedly behaved badly to her family, and she said with astonishment: 'Who would have believed it of such a charming voice?' When she heard singers, she distinguished between *dark* and *fair* voices.

When people spoke to her, she judged of their height by the direction of the sound, which came to her from above if the person speaking were tall, and from below if that person was short.

She was not anxious to see, and one day I asked her the reason. 'The reason,' she replied, 'is that then I should only have my own eyes, whereas now I have the use of everybody's; by this loss I am always an object of interest and pity, at every instant people do me kindnesses, and at every instant I am grateful Alas! if I could see, no one would trouble about me.'

The errors to which sight is liable diminished its value in her eyes. 'I stand,' she said, 'at the entrance of a long alley; and there is a certain object at its far end. One of my friends sees it moving, another sees it stationary; one says it is an animal, another that it is a man; and on approaching it, it turns out a tree-stump. No one can tell if the tower they see in the distance is round or square. I brave a whirlwind of dust, while those about me close their eyes and become ill–sometimes for a whole day – because they had not shut their eyes soon enough. An imperceptible atom is enough to cause them cruel pain.' At the approach of night she used to say that 'our reign was drawing to a close, while hers was beginning.' Living in the dark, and accustomed to act and think during this eternal night, insomnia, which is such a burden to us, had no terrors for her.

She would not forgive me for my statement that the blind, to whom the symptoms of suffering are invisible, must be cruel. 'Do you imagine,' said she, 'that you hear a cry of pain as I do?' 'There are people,' said I, 'who suffer in silence.' 'I believe,' she said, 'that I should soon discover them and pity them all the more.'

She was devoted to reading and passionately fond of music. 'I think,' said she, 'I should never tire of hearing good singing or playing; and if this were the only pleasure in heaven, I should not be sorry to go there. You are right in maintaining that music is the most impassioned of the fine arts, not excepting poetry and oratory; that even Racine does not express himself as subtly as a harp, that his music is heavy and monotonous when compared with an instrument, and that you have often wished to give your style the strength and lightness of Bach's music. Music is the most beautiful language I know. In spoken language, the better we pronounce words, the more particularly we articulate each syllable; whereas in the language of music, sounds of the most widely different pitch from bass to treble and treble to bass follow one another imperceptibly, forming one single prolonged syllable, which varies its inflexion and expression at every moment. While this syllable is brought to my ear by the melody, the harmony carries it out, without any confusion, upon a number of instruments – two, three, four, or five perhaps – which all combine to strengthen the expression of the melody. I can understand the music without the words sung, if the symphonist is a man of genius whose music is full of character and expression. Music is most delicious and expressive in the silence of night.

'I fancy that people who see, distracted by their eyes, cannot listen and hear as I can. Why does the praise of music I hear seem poor and faint? Why can I never speak of it as I feel? Why do I pause in the midst of what I am saying, seeking vainly for words expressive of what I feel,? Are such words not invented? I know nothing comparable to the effect of music but the joy I feel when, after a long absence, I throw myself into my mother's arms; my limbs tremble, my tears flow, and my knees totter, and I feel as if I should die of joy.'

She had the most delicate feelings of modesty; and when I asked her reason, she replied: 'It is the result of my mother's teaching, who has so often told me that the sight of certain parts of the body is an invitation to vice. I confess I have only understood her lately, and perhaps I had to become less innocent to do so.' She died of an internal tumour which she never had the courage to inform anyone about.

She was extremely neat and clean in her clothes and person, and this is the more remarkable as she had not eyesight to assure her that she had been successful in avoiding the vice of uncleanness and untidiness.

When her glass was being filled, she knew by the sound of the liquid as it fell, when it was full. She fed herself with surprising dexterity. Sometimes she amused herself by standing before a mirror to dress herself, and by imitating all the affectations of a coquette. The mimicry was so true to life that we laughed aloud.

From her earliest youth efforts had been made to train her other senses, and the results were astonishing. Touch enabled her to discern minute details in shapes of objects which often pass unnoticed by those who have the best eyesight.

She had very delicate senses of hearing and smell; she knew by the feeling of the air whether the weather was cloudy or fine, whether she was walking in a square or a road, in a road or a cul-de-sac, in an enclosed or open place, in a vast apartment or a small room. She measured the space by the sound of footsteps or the echo of voices. When she had gone over a house, its plan remained in her head, so that she would warn others of little obstacles or dangers in their way. 'Take care,' she would say, 'the doorway here is low; you will find a step there.'

She noticed a variety in voices which we have no conception of, and when she had heard a person speak once or twice she knew him for ever.

She was very little affected by the charms of youth and by the wrinkles of age; and said she was only charmed by the fine qualities of the heart and intellect – one of the advantages of the loss of sight, especially for women. 'My head will never,' she said, 'be turned by a handsome face.' She had a very trusting disposition. It was so easy, and would have been so shameful, to deceive her. To lead her to imagine she was alone in a room, when this was not so, would have been the blackest of treacheries.

She was never subject to panic, and rarely to ennui; for she had learnt in her solitude to be independent of others. She noticed that at nightfall in travelling in public vehicles people became silent. 'I do not need,' she said, 'to see those whom I love to converse with. 'She set the greatest value upon sound judgment, sweetness of disposition, and gaiety. She spoke little, and was an excellent listener. 'I am like the birds,' she said: 'I learn to sing in the dark.'

When she compared what she heard from day to day, she was astonished at the contradictory nature of our opinions; praise or blame seemed to her a matter of indifference from such inconsistent creatures as human beings.

She had been taught to read by cut-out letters. She had a pleasant voice, and sang with taste, and would have gladly spent her life at concerts or operas; the only music she did not care for was noisy music. She danced exquisitely, and also played the viol very well, and owing to this talent she was greatly in demand among young persons of her age, to whom she taught the fashionable dances.

She was the best beloved of her brothers and sisters. 'You see,' she said, 'what I owe to my infirmities; people become attached to me as a result of their kindness to me, and of my efforts to show my gratitude and deserve their good offices. Besides, my brothers and sisters are not jealous. If I had sight, my heart and intellect would be the losers. I have so many inducements to be good! What would become of me if I were to lose the interest that I inspire?'

In her parents' loss of fortune, the only thing she regretted was the loss of her masters; but they had so much liking and esteem for her that her music and mathematical masters begged her to let them teach her for nothing. She asked her mother: 'Mother, what am I to do? They are not rich, and need all their time.'

She had been taught music by notes in relief placed on raised lines on a large board. She read these notes with her hand, and played them on her instrument, and in a very short time she learnt to play the longest and most elaborate piece.

She knew the elements of astronomy, algebra, and geometry. Her mother, who read her Abbé de la Caille's book, would now and again ask her if she understood it. 'Quite easily,' she would reply.

She declared that geometry was the science of the blind, because it was of such universal application and no external aid was necessary to become proficient in it. 'The geometrician,' she added, 'spends nearly all his life with his eyes shut.'

I have seen the maps with which she studied geography. The parallels and meridians were made of wire; the boundaries of kingdoms and provinces of embroidery in linen, silk, or wool of various thickness; the rivers and streams and mountains of pins' heads of various sizes; and cities and towns of drops of wax of various sizes.

One day I said to her, 'Mademoiselle, imagine a cube.'

'I see it. '

'Place a point in the centre of the cube.'

'I have done so. '

'From the point draw straight lines to the angles; into what have you divided the cube?'

'Into six pyramids,' she replied without hesitation, 'each having as its base one side of the cube, and a height equal to half its height.'

'True, but tell me where you see this?'

'In my head, as you do.'

I must admit I have never been able clearly to understand how she represented figures in her head without the aid of colour. Was her cube formed from memories of sensations of touch? Had her brain become, as it were, a hand within which substances were realised? Had a connection between two senses been established? Why does this connection not exist in my case, and why do I picture nothing that is not coloured in my mind's eye? What is the imagination of a blind man? This phenomenon is by no means easy of explanation.

She wrote with a pen, with which she pricked a sheet of paper stretched on a frame divided by two parallel and movable slats, which only left sufficient space between them for one line of writing. The same method of writing served to answer her, as she read the communication by passing her finger-tips over the slight roughness formed on the back of the paper by the needle or pin.

She read books printed on one side of the paper only for her use by Prault. One of her letters was printed in the *Mercure*.

She took the trouble to copy out with her needle President Hénault's *Historical Synopsis*,[25] and her mother, Madame de Blacy, gave me this curious document.

People will find it difficult to accept the following fact, though I and all her family, as well as twenty persons still alive, can vouch for it. Given a piece of poetry of twelve to fifteen lines, if she was told the first letter and the number of letters in each word, she could reconstruct the poem, however odd and far-fetched. I tried her with Colle's[26] ambigouris. She sometimes lighted on a better word than the original. She threaded the finest needle rapidly by laying the thread or silk on the index finger of her left hand and drawing this with a fine point through the eye of the needle placed perpendicularly. She could make all sorts of small articles – edgings, bags of all kinds, some of drawn work, and of various patterns and colours; garters, bracelets, necklaces made of glass beads the size of letters arranged to form patterns. I am sure she would have made a good compositor, for the greater includes the less.

She played reversis, *médiateur,* and quadrille well. She sorted her cards herself, and recognised each by touch from minute peculiarities others could neither see nor feel. In reversis she had a special place for the ace (especially the ace of diamonds) and the knave of hearts. The only difference in playing with her was that the card played was named. If the knave of hearts was in danger, a smile passed over her face, which she could not restrain though she realised that it was indiscreet.

She was a fatalist, and believed that our efforts to escape our destiny only served to draw us closer to it. I do not know what she thought of religion; she kept her opinions to herself out of consideration for her mother, who was devout.

Lastly, I will give you her ideas upon handwriting, drawing, engraving, and painting, and they are, I think, very just, as I hope you will think after reading the following conversation between us. She begins the dialogue:

'If you trace on my hand with a point, a nose, a mouth, a man, a woman, or a tree, I should be sure to recognise them; and if the tracing was correct, I should hope to recognise the person you had drawn; my hand would become a sensitive mirror, but the difference in sensibility between this hand and the organ of sight is immense. I suppose the eye is a living canvas of infinite delicacy; the air strikes the object, and is reflected back from the object to the eye, which receives a multitude of impressions varying in accordance with the nature, the form, the colour of the object, and also perhaps with the properties of the air which I do not know, and of which you are equally ignorant; and the object is represented to you by the variety of these sensations.

'If the skin of my hand was as sensitive as your eye, I should see with my hand as you see with your eyes; and I sometimes imagine there are animals who have no eyes, but can nevertheless see.'

'And the mirror?'

'If any bodies are not mirrors, it is by some defect in their composition which destroys the reflection of the air. I think this is the more likely as gold, silver, iron, and copper, when polished, are able to reflect the air, while rough water and cracked ice lose this property. Variety in sensation (and hence in the property of reflecting air), in the materials you employ, distinguishes the writing from the drawing, the drawing from the engraving, the engraving from the picture. The writing, the drawing, the engraving, and the picture in one colour are all monochromes.'

'But if there is only one colour, we should only distinguish that colour.'

'It seems that the surface of the canvas, the depth of colour, and the way in which it is used, produce in the reflection of the air a corresponding variation to that of the objects. Don't ask me any more, for that is all I know.'

'To try to teach you any more would be waste of time.'

I have not described in her case all I might have noticed if I had seen her oftener and questioned her skilfully. I give you my word of honour that all I have recorded is actual fact.

She died at the age of twenty-two. With a wonderful memory, and strength of mind as wonderful, what progress she would have made in science if she had had a longer life! Her mother read history to her, and this task was pleasant and useful to both of them.

Notes

1 The *Letter* was addressed to Madame de Puisieux. – (A)
2 The original is; *Possunt quia possunt videntur* [They succeed because they think they will succeed].
3 Hilmer, a Prussian ocultist. – (Br)
4 A small town in the Gàtinais. – (D)

5 Lieutenant of police. – (Br)

6 Clement (*Cinq années littéraires*, lettre xxxiii) chooses this passage to give his correspondent some idea of this new book of Diderot's which he describes as obscure, and in which he only finds a very slight exhibition of learning. – (A)

7 Brière gives the name of this geometrician as Rapson (*sic*). Raphson, not a very distinguished mathematician, may, among many others, have quoted this doctrine of Plato, but it is not very important if he did so. What makes the dictum important in Plato's mouth is that he had a theory that geometry is more fundamental and comprehensive than arithmetic. He disagreed in this respect from the Pythagoreans because he clearly realised that there were certain lengths of lines expressible geometrically but not arithmetically; cf. Brunschvieg, *Les étapes de la philosophie mathématique*, pp. 45, 47, 48.

8 See note 1, p. 000.

9 Printed in London, a year after Saunderson's death, at the expense of Cambridge University. In 1756 de Joncourt translated it, with some additional remarks (Amsterdam, 2 vols.). – (Br)

10 Naigeon, and after him the editor of 1818, have inserted, instead of the initials *M. de M* ... in the original edition, M. de Montesquieu. This is a great mistake; Diderot himself has given M. de Marivaux in the index of the 1749 and 1751 editions. The *Esprit des Lois* appeared in 1748, which might have caused this error on the part of the editors, who had not consulted the index. – (Br)

11 *Dialogues between Hylas and Philonoüs* (1713), translated by the Abbé Gua de Malvin (1750). – (A)

12 Condillac (1711–1780), whose *Essay on the Origin of Human Knowledge* had appeared anonymously in 1746. – (A)

13 Memoirs of the life and character of Dr Nicholas Saunderson in Saunderson's *Algebra*, vol. i, p. xi (1740).

14 See note 2, pp. 219, 220.

15 See Introduction, pp. 10–15, 18, 19.

16 See note 3, pp. 221, 222.

17 This is the thesis of Lucretius, and the theory of the survival of the fittest. – (A)

18 By rendering a Dr Inchlif responsible for his imaginary reconstruction of Saunderson's last moments Diderot alienated the sympathies of England. – (A)

19 See note 4, pp. 222, 223.

20 See note 5, pp. 223–5.

21 [This translation has been collated with an eighteenth-century translation, undated and anonymous, entitled a *Letter on Blindness*. It has further been amended for the 1999 Clinamen Press edition.]

22 'We have appended to the *Letter on the Blind* the sequel which Diderot composed a long time after it.... Those who accuse the writer of having always written hastily or of having always been hard and positive have certainly not read all his works. This sequel alone would confute them.' – (*Depping, B*)

23 'You would be wise to turn loose the old horse in good time, lest he fail in the end, amid laughter and strain himself.'

24 Jacques Daviel, surgeon, born in 1696. In 1728 he made a special study of diseases of the eye, and acquired such a high reputation for skill that in the month of November 1752 alone he performed two hundred and twenty-six operations for cataract, of which one hundred and eighty-two were successful. He died in 1762. – (A)

25 *Abrégé de l'histroire de France.*

26 Collé, Charles (1709–1783), a dramatic author who also wrote ambigouris or nonsense verses.